John Nelson Smith

On the Science of Sensibility (Intelligence)

Or Simple Element of Soul and the Spirit of Life and Origin of Species

John Nelson Smith

On the Science of Sensibility (Intelligence)
Or Simple Element of Soul and the Spirit of Life and Origin of Species

ISBN/EAN: 9783337024062

Printed in Europe, USA, Canada, Australia, Japan

Cover: Foto ©berggeist007 / pixelio.de

More available books at **www.hansebooks.com**

ON THE

SCIENCE OF SENSIBILITY,

(INTELLIGENCE,)

OR

SIMPLE ELEMENT OF SOUL;

AND THE

SPIRIT OF LIFE AND ORIGIN OF SPECIES,

AND

NATURAL CAUSE OF THE CONSTANCY OF EACH SPECIES TO ITS TYPE.

BY JOHN NELSON SMITH.

PHILADELPHIA:
J. B. LIPPINCOTT & CO.
1875.

Entered according to Act of Congress, in the year 1874, by
JOHN NELSON SMITH,
In the Office of the Librarian of Congress, at Washington.

TO

ULYSSES S. GRANT,

PRESIDENT OF THE UNITED STATES,

WITH APPRECIATION OF THE INDOMITABLE SOLDIER,

THE CAUTIOUS STATESMAN, AND THE MODEST CITIZEN,

THIS WORK IS

RESPECTFULLY DEDICATED

BY THE AUTHOR.

CONTENTS.

	Page.
INTRODUCTION	9

CHAPTER I.
FORMS AND SOCIAL HABITS OF SUNDRY TYPES AND RACES.... 40

CHAPTER II.
THE SOUL—ITS STRUCTURE AND MOVEMENT...................... 49

CHAPTER III.
ELEMENTARY PRINCIPLES OF SOUL.................................. 57

CHAPTER IV.
EMINENT RELATIONS BETWEEN SUBSTANCE AND FORCE........ 65

CHAPTER V.
DEPARTMENT OF SOUND.. 75

CHAPTER VI.
DEPARTMENT OF LIGHT... 80

CHAPTER VII.
FINITE AND INFINITE.. 85

CHAPTER VIII.

REMARKS ON INFINITUDE.. 103

CHAPTER IX.

SIMPLE ELEMENTS OF SOUL, SPIRIT OF INTELLIGENCE......... 112

CHAPTER X.

SIMPLE ELEMENTS OF THE FORCE OF ANIMAL GROWTH........ 128

CHAPTER XI.

COMBINATION OF ELEMENTARY FORCE OF GROWTH EMPLOYED IN THE PRODUCTION OF SHELL-FISH................................ 134

CHAPTER XII.

PISCATORY AND SNACAN FORCES OF ANIMAL GROWTH.......... 142

CHAPTER XIII.

EMBRACING THE FORCES OF ANIMAL GROWTH EMPLOYED IN THE PRODUCTION OF WORMS AND INSECTS, CALLED ARTICULATES... 152

CHAPTER XIV.

FOWLS 158

CHAPTER XV.

ELEMENTS OF FORCE OF ANIMAL GROWTH—QUADRUPEDS..... 175

CHAPTER XVI.

FORCE OF GROWTH, AS ARRANGED FOR THE PRODUCTION OF BIPEDS.. 198

INTRODUCTION.

In the books which have come under our observation we note three particular references to the construction, variety, and definition of soul: One by Moses, one by Solomon, and one by Webster, each of which we shall refer to in its proper order.

Webster defines it to be "the spiritual, rational, and immortal part in man—life, vital principle; that part of man which enables him to think, and which renders him a subject of moral government; —sometimes, in distinction from the higher nature or spirit of man, the so-called animal soul, that is, the seat of life, the sensitive affections and fantasy, exclusive of the voluntary and rational power;—sometimes, in distinction from the mind, the moral and emotional part of man's nature, the seat of feeling, in distinction from intellect;—the understanding; the seat of knowledge, as distinguished from feeling."

Thus Webster makes soul the embodiment, essence, warp and woof, of that which gives vitality to man: motive power, cause of development, reasoning capacity, ideas, judgment, memory, and determines the course of his actions, directs his mode of life, and constitutes him a self-acting institution.

If soul does this for man, what does the same

thing for other animals, even fowls, fish, and monads? Yea, verily, every living creature that moveth on the surface of the earth or in the water is a self-moving, self-sustaining institution, susceptible of pain; enjoys pleasure; apparently reasons from cause to effect; are sometimes in a good humor, sometimes very angry; play, fight, and roam where they list; come at the call of man, and go at his bidding; and have ideas, judgment, and memory; propagate their species by pairs through copulation; are nourished through infancy by their parents, till they are capable of seeking their own food, when they go out and provide for themselves, till the body is rendered useless by age, disease, or vital injury, when they die:—and what more does man do?

Still, we are gravely told by sages, theologians, parents, and teachers, that one is the work of an immortal soul, and the other of a mortal instinct. What is instinct?

One author defines it to be, "urged or stimulated from within to do what the most perfect reason would do; moved, animated, excited."

If instinct does what the most perfect reason would do, why not call it reason? If that principle which causes a beast to perform its work of life with greater perfection than man does his is mortal, why is that principle which causes man to perform his work less perfect immortal?

If the work of a beast, which is voluntarily

prompted by a will produced in his own mind by a desire of gain, which his judgment tells him will supply his wants, is instinct, by what other process is the work of man performed?

Does man have any instructor but his own soul, any teacher but the spirit which God gave him? Did not he commence his work of civilization just as voluntarily as the beaver builds his dam and substantial habitation, the fowl its nest, or the bee makes honey?

If one is instinct, why is not the other? If one is reason, why should not the other be called by the same name; as the only visible difference is that each species of animal does a work peculiar to itself, and is confined to a circle of greater or less degrees, and each enjoys itself in a mode of life more or less disgusting to every other.

As man is begotten and born, so are all other animals, and the vital functions of life are all performed in the same mysterious manner, and each supplies itself with food through life, and all die in the same manner: then why should one be immortal, and all the others mortal; one prompted by reason, and the others by instinct?

Oh, says the reader, you ask so many questions all in a string, that they become very perplexing, and it is doubtful whether you can answer them all yourself satisfactorily.

Yes, we are aware that these have, in times past, been very perplexing questions: not so much from their abstruseness as from the imperfections and follies of man himself—even that pertness which causes him to rise to the surface, like a gas bubble, and pop off in a squib of ridicule at everything which he does not fully understand;—that pompous, don't-come-near-me feeling which prompts the expression, "stand off, I am more holy than thou;"—that exclusiveness which secretes him in a hermit's cell and isolates the paterfamilias in the midst of a city;—that clannish spirit which prevents the Jew from marrying with any but a true descendant of Israel, the Orangeman from associating with a Catholic, the lord with a peasant, or pagan with Christian;—that fanaticism which leads every sectarian to believe that his Church is the only door through which any man can possibly enter heaven;—that blind zeal which convinces every bigot that the political institutions under which he lives are the best form of government that can be instituted among men;—

In short, that microscopic view of things in general, that ignorance of surrounding elements, which caused the universal belief only a few centuries ago that the earth was flat and stationary as a house on a plain, and that the sun, moon, and stars all moved around it; the Chinese, that they are celestials, and all the other nations are barbarians; the Jews, that they are the favored people of God, and all other people rejected Gentiles; Christians certain that pagans, Jews, and Mohammedans are all doomed to eternal burning; and has caused the whole family of man to consign every other species of animal to annihilation.

Each individual, confining his observation to a superficial comparison, exclaims, the whole animal kingdom was made for my use, ("see man for mine, replies a pampered goose;") yes, all of you are created expressly to prepare milk and honey and meat for my table, (see man for ours, replies a pack of wolves,) material for my clothing, and a propelling power for my person, (see man for mine, says a flea.)

You groveling beasts have no souls, minds, or sensibility, and your ideas I will beat out of you, unless they correspond exactly with mine. Your wants and enjoyments are not entitled to the least consideration.

You are my tools, which I will use at my convenience, and then expose you to cold and starvation. You are my machine, that I will use to till and reap and fill my garners, and then let you die of starvation in the winter.

Shiver in the pelting storm, do you? Don't tell me you suffer. Don't look piteously about my garners, which you have rendered such good service in filling. Don't tell me that hunger gnaws you, or the cutting wind drives a pang to thy mind. I don't believe you can suffer. You are only a kind of self-propelling V-I-T-A-L— an instinct, that's what you are—animal instinct;

a poor dumb brute. You have no soul. I will let you live as long as you will work for me and feed upon the wind. I am man, the lord and proprietor of the earth and all that is on it, (and I of the plains and all that is on them, says the lion, and seizes the boasting hero, and bears his mangled carcass away to his den, and divides it among his whelps.)

Many even doubt the existence of a God, and consider themselves the highest order of intelligences; look upon the laws of nature as an incomprehensible self-existent dummy, without sensibility; and so-called philosophers and theologians have invented myriads of absurd theories on every subject, which are constantly leading their minds and pursuits away from the truths which, if understood, would contribute most to man's felicity.

Is it strange then that, amid this fog of perplexity concerning all terrestrial things, the phenomenon of soul should be a very perplexing question, and but little understood? Even Solomon, who by common consent has been placed at the head of the wise men of his age—a potent king, sagacious judge, and instructive preacher—was so perplexed by it that he finally pushed through the fog, made a thorough examination of it, and inscribed the following sage opinion in the poetic language of the Preacher. (Ecclesiastes, iii : 17–22:)

"I said in my heart, God shall judge the

righteous and the wicked: for *there is* a time there for every purpose and for every work.

"I said in my heart concerning the estate of the sons of men, that God might manifest them, and that they might see that they themselves are beasts.

"For that which befalleth the sons of men befalleth beasts; even one thing befalleth them: as the one dieth, so dieth the other; yea, they have all one breath; so that a man hath no pre-eminence above a beast: for all *is* vanity.

"All go unto one place; all are of the dust, and all turn to dust again.

"Who knoweth the spirit of man that goeth upward, and the spirit of the beast that goeth downward to the earth?

"Wherefore I perceive that *there is* nothing better, than that a man should rejoice in his own works; for that *is* his portion: for who shall bring him to see what shall be after him?"

We think if Solomon had transposed that poetic effusion, and written it out himself in plain prose, it would have run in this wise:

I have considered the phenomenon of animal life, and have discovered that the moving principle of each living creature is an immortal spirit, forever retaining its personal identity, all of the same species being of the same quality. Consequently they adopt the same habit of life, cleave to each other for social enjoyment, and propagate a holy seed, so that their generations roll on unamalgamated, each performing its own work in perpetual routine, each having a certain power of action, by which its movements are directed and habits of life formed, fixed, and determined. Therefore let man rejoice in the work of his own hand, for that is his portion, and glorify God, with the great capacity which he has given him, in as subordinate a manner as other animals do with their small capacity.

But, says the reader, what difference does it make to me whether a beast has a soul or not? It won't add to my pelf, acres, or garner. I don't see anything in it that is going to pay me for the trouble of the investigation. We answer, go up with us through that opening in the fog which Solomon's truth hath made, keep your eyes and ears open, and you will discover that the most portentous storm is now gathering in the clouds that lower about this very question which has ever burst upon the world since the Deluge; and unless the lusts of mankind and present tendencies of political events are changed, we shall soon have no pelf to add to, no garners to fill, no civilization to rejoice in; and that a thorough understanding of this question is the most important subject for consideration of this political decade.

But in attempting to work our way through the cobwebs of false theories and fog of skepticism into the sunny regions of truth, which the truth of Solomon gave a faint glimpse of, the first obstacle we meet with is an opinion, prevailing to a considerable extent, that the idea of soul is a delusion—a will-o'-the-wisp—that exists only in imagination.

Having discovered the fact that man is only the head link in the great chain of animal life, rather than stem the current of universal unbelief in the immortality of beasts, timorous investigators give up their own immortality rather than attempt to maintain the immortality of other animals, which manifest the same evidence of it that man does, and so drop the whole mass of animal life into oblivion: and still those same men know and acknowledge that no particle of any substance whatever can be annihilated, whether found in the materials which compose the earth, water, or gases; electricity, light, or polar attraction; gravity, crystallization, centripetal or centrifugal forces—yea, verily, that no substance, ponderable or imponderable, can ever change its nature in the slightest degree; and that whatever does exist in any inanimate form is eternal. Still they tell us that that principle, which is the prompter of sensibility, the constructor of all vegetable specimens, and the maker of every specimen of animate carcass—the only thing in nature which is endowed with intelligence, and is capable of appreciating the splendor of colors, harmony of sound, and of rejoicing in its own existence, and of signing hosannas to its creator—is only an imaginary phantom: in fact has no existence at all. And still they know that without it there could be no imagination of a phantom or anything else. If its existence is acknowledged at all, then the proof that it can cease to exist devolves upon them, which is impossible.

Some master-minds become so elated with their discoveries in the arts and sciences, that, looking down from the eminence to which they have climbed over the mass of mankind and other animals, which at so great a height seem an indiscriminate mass of animal life, between which they can make no immortal distinction, they come to the conclusion that they are all a mortal set of muck-worms—mere animated dust. Then, raising their eyes and gazing around them upward and downward for God, but cannot see him,—stretch out their arms into space, and hook and claw with their fingers, but cannot feel him,—they come to the conclusion there is none: like the great Compte, who, after elaborating a masterly analysis of the sciences, and placing himself at the head of the scientific men of France, exclaimed, "I can find no spirituality of the universe; no soul in man."

Still, show him a well-regulated watch, and he would swear it had a maker; a moving steamship, and he will protest some man constructed it and superintended every movement. Bring him into a telegraph office, and he will say the messages are conveyed by the galvanic force, in accordance with the will of man, which he knows no more about than he does about the attributes of Deity.

The inventor of the telegraph he has never seen, nor the builder of the ship, nor the maker of the watch, neither does he know the time or mode of their construction; would not pretend to say that he could make either of them: still he believes and acknowledges that they had an intelligent maker.

But when he looks at the heavenly bodies, and sees their magnitude and wonderful movement, in the most perfect harmony, throughout the immensity of space,—each planet careering through its orbit and accomplishing its destined circuit with a precision so nicely adjusted that an astronomer can calculate the exact time and size of an eclipse for years ahead, and the whole grand machinery of the planetary system, spread out through infinite space, all governed and controlled by one infinite mind in a harmonious, uniform movement, without variation or shade of change in procession, between which and any machinery which man has ever organized the comparison sinks man's best effort to a rickety mechanical abortion, an unseemly wreck of discord, mostly out of order,— a finite machinery of disappointment, disaster, and death to its maker and those who use it,—he exclaims, "All chance; no design in its arrangement; no intelligent condition of its movement; no architect of its sublime plan; no builder of the structure by the prepared model!"

But in the getting up of the steamship Great Eastern he sees an architect so eminent that a monument should be built to his memory, and builders so skillful that they are worthy of all praise and honorable medals; and he is ready to bend the knee and bow in homage to the substantial men of capital who had the courage to furnish the vast sums of money required in the experiment.

But when asked to bow in homage to the infinite architect who planned and constructed the universe, and controls the whole planetary system with a fiat of harmonious order that is never broken, he shrugs his shoulders, elevates his eyebrows, and with curled lip sarcastically answers, "Sir, I don't see him; an infinite intelligence of omnipotent creative powers is impossible. The intelligence which I see about me is all that I can possibly believe in, and that extends not beyond the dissolution of this body. Death makes an end of joy and sorrow, pain and pleasure; indeed, sir, the sun, moon, and stars must have constructed themselves, and marched into mechanical order without any rational design, and still pursue their routine of movement without any directing hand;—all blind chance."

What inconsistency! Why should not the steamship Great Eastern have constructed itself, and taken up its transit between Liverpool and New York for the benefit of commerce by the same chance—know-nothingism—that the moon created itself and took up its regular circuit around the earth for the benefit of night pedestrians? This egotistic so-called scientific professor never saw the architect of the steamship Great Eastern; still he believes in him because he has seen his works, but denies the existence of the Infinite Architect who orders all the courses of nature, and of whose creative genius he himself is a living demonstration, and cannot open his eyes without seeing myriads of his works.

Oh, vain man, who to make a god of himself would dethrone Deity, and ignorantly proclaims that what he cannot make, made itself by a chance blunder of know-nothingism!

Another stumbling-block which has been rolled into the straight and narrow path of truth is intelligibly condensed and presented to the mind in the following stanzas of a poem by J. G. Severance, Esq., of Sacramento, California:

> "Progressing still, the latter, fecund earth,
> Brought forth the sterile plants of baby birth,
> As incubated eggs produce; and these
> Bred grass and herbage, flowering shrubs and trees,
> With trunks and limbs, and leafy lungs for breath,
> The vital heart, and germs of life and death.
> There is a plant, the botanists have found,
> Which, though adhering to its parent ground

> More perfect than mimosa, yet seems rife
> With a distinctive, animated life:
> In the progressive chain, this is the link
> Connecting earth with particles that think;
> For slight advancement makes these plants to be
> Perambulating animalculæ.
> These turn to insects, they to reptiles grow,
> Still groveling on the earth, with movement slow,
> Till time's improvement lifts them from the sod
> And makes them higher, nobler, more like God.
> Instinctive beasts then follow in their course,
> The lower order first, and then the horse
> And dog, sagacious, capable of thought,
> Of hate and friendship, and of being taught;
> The monkey then, half human and half beast,
> Then they, of all who walk upright the best,
> A race with tails, on Afric's southern coast,
> O'er whom the negro slight advance can boast;
> On through the Indian we the lineage trace,
> Of the reflective, pure, Caucasian race."

Thus these metaphysical paradox weavers all admit, in some one form or another, that this ignorant form of nature was not content in his ignorance, but has been eternally striving to manifest some degree of intelligence, in order that he might have a little joy in his existence, and for this purpose, at some unknown period, millions of millions of years ago, developed himself into an intelligent plant with a sensible blossom, out of which sprouted a worm, and that in that form he crawled and wriggled and transmuted for a million or two of years, till he finally developed himself into the highest specimen of a serpentine boa constrictor, which actually knew enough to lie in ambush when it was hungry, and seize some victim of less dimensions than itself and devour it for food.

Thus their god of infinite ignorance crawled and wriggled in the form of worms, snakes, lizards, terrapins, and alligators for ever so many millions of years, till he finally manifested the intelligence of a four-legged beast, when he fed upon grass and herbage for millions of years, during which time he transmuted and changed till he finally got a sufficient amount of jaw-bone, tusks, and grinders, and strength of muscle to enable him to commence feeding upon himself in a more groveling state, by which means he was able to make greater progress, and in a few millions of years actually developed into the form and with the intelligence of a devouring lion, tiger, and a multitude of cannibal beasts.

Not content with that development of irresistible animal strength and devouring intelligence, he determined to part with a portion of his legs and muscle, and strive to compensate himself therefor by an increase of brain and intellect. So he tried another transmutation, and in the course of a million or two of years made his bow to the world in the form of a monkey, with a tolerable pair of hands. And in the course of a billion or two of succeeding years' diligent striving, by sundry transmutations and patient groveling in the form of baboons, gorillas, orang outangs, chimpanzees, bushmen, caudal negroes and negroes proper, Esquimaux, Navajo and Digger Indians, he at last proclaimed himself sovereign ruler of himself, as demonstrated in every other form of the animal kingdom, in the form of a man, with sufficient intelligence, tact, and strategy to devour them all, even the lion; but had no more idea of the laws of nature or the manner in which he had created all things—not even himself in the form of a man—than a tree or vegetable. He even believed for millions of years that the earth was flat and stationary, and that the whole planetary system revolved around it: and this, we are gravely told, is the highest order of intelligence that God has been able to manifest himself in.

What a god to worship. What a strange jumble of inconsistent contradictions. The creator of all things, in the very highest condition of intelligence and broadest sphere of knowledge of the past, present, and future in which he has ever been able to manifest himself, has not the least idea of how all things were created! Notwithstanding he himself has done it, he has not the slightest knowledge of model, mode, or means. Still these dogmatists tell us that the only intelligence which God hath is demonstrated in the doings of this infinite multitude of cannibal beasts, of which man is chief: for he kills for the pleasure of killing, and leaves the carcasses to rot and stink on the face of the earth out of pure deviltry, while every other carnivcrous beast kills only for food; and when his hunger is satisfied, he dwells in peace with them till hunger demands another victim.

The above theory had its origin in the minds of men assuming to be gods at a very early day in the doings of the family of man; for the very first pages of African and European history inform us that it was the religion of the ruling dynasties of both these continents, "the Israelites excepted," down to the decade when Constantine became a convert to Christianity, and proclaimed to the world his belief in the personal, omnipotent God of high heaven, who through infinite wisdom had created all things, and controls universal nature by his irresistible fiat, and that he also believed that Christ Jesus of Nazareth was his only begotten son, before whose name every ruling dynasty tumbled down and broke off their pagan head of graven image devotion, as Dagon fell down before the ark of the covenant of God which Moses had made. So that for over a thousand years the ruling dynasties of both these continents have been worshipers of Abraham's God, have prayed to him as Abraham prayed to him, appointed thanksgiving days and given thanks as David, the sacred poet of Israel, did, and the daily sacrifice to a beastly devouring god has long been abolished.

Much prejudice exists among so-called Christians against Mohammed. But his declaration stands recorded in the Koran in the plainest and most positive language, that Christ was the begotten Son of God by an immaculate conception, and all of his followers are the most devout worshipers of Abraham's God, and they themselves are the seed of Abraham.

Abraham's idea of God was identical with that faith which instigated Noah to build the ark and made a preacher of righteousness of him for three hundred years after the flood, from whose own lips Abraham received the sacred truths: for the records show that Abraham was fifty-six years old when Noah died, and during that three hundred years Noah fixed the same faith and idea of the same personal, intelligent, Holy God, of infinite goodness, so firmly in the posterity of Shem, who settled Asia, that their ruling dynasties have never departed from it; which is the cause of the stability of their government. So that for over a thousand years the entire families, kindreds, tongues, and nations of men have been governed by ruling dynasties who believed in the God of Noah and of Abraham.

When Ham, whose posterity peopled Africa, and Japheth, whose posterity peopled Europe, conspired and rebelled against the rule and government of their father and elder brother, they rebelled also against their God, and denounced the high God of heaven, and proclaimed an endless sec of beastly deity, "which lives through all life, extends through all extent, spreads undivided, operates unspent," the highest order of intelligence discernible in which was that of a devouring beast, which they called father; and as they believed granite was the base of earthly formation, that was the matrix in which they had been begotten and brought forth, as it is described in the above poem, by a vegetating, crawling, wriggling, serpentine process; hence the worship of granite images of serpents, beasts, and apes in the olden time. Even Alexander claimed to have been begotten by the god of war Jupiter, in the form of a serpent, which was seen to glide from his mother's bed about the time of his conception. Thus those pagan worshipers of a serpentine, devouring, beastly god of war, carnage, and devastation, claimed it as their father and granite as their mother.

Under this pagan hallucination the sacrificing of men, women, and children, to appease the anger of those terrible gods, in the most excruciating, diabolical manner, repugnant to every human sympathy, was practiced. They even caused their children, in groups, to pass through fire of such intense heat that many died in the flames, and all suffered the most intense torture, to appease the god of fire, that their houses and crops might not be burned.

That terrible superstition dragged them down to the darkest ages of barbarism, in which every man who had sufficient family influence and personal stamina to form a ring set himself up as king, and war, plunder, and devastation were the order of the day; and even as early as the epoch when Joshua established the nationality of the Israelites in the land of Canaan, there were as many petty kingdoms in it as there are swarms of bees in an apiary, all of which were in perpetual warfare with each other, and each had selected some one of the beastly deities for their god, and went to war and battled in his name, and when they were defeated, attributed it to his anger, and sacrificed their children to appease him. So also when the Philistines took the ark of the covenant, the supposed sanctuary of Israel's God, from them in battle, they attributed it to the mastery of their god over Israel's God, and carried it home and set it in the temple of their god Dagon, that he might rejoice over his conquered foe in his own temple. But the next morning their priest, on entering the temple to worship, found Dagon prostrate before the ark, and gave the alarm in great consternation, and Dagon was again set up and propped with extra braces, and behold, the following morning he had fallen again, and broken off his head across the door sill, and in great consternation they sent the ark of the covenant of the people of Israel with their God, even the sanctuary which contained the tables of stone on which God himself had written the law by which they executed judgment among the people, back to them, lest he should destroy them as he had their god.

We presume there is not a reader of this work but does believe and will acknowledge that there is a force in nature that can shake the Sierra Nevada mountains, with half the American continent, till the inhabitants thereof quake with fear—cities shaken down, hamlets destroyed, and thousands of people buried in the ruins. But when they read the above paragraph of Jewish history, that the controller of said force did shake down the granite graven image of Dagon, they shrug their shoulders and cry bosh, only a Jewish fable! Why should God put himself to the trouble of shaking down a heathen idol? Answer: For the same reason that he should trouble himself to destroy the ruling dynasties who worshiped those devouring beastly gods which the graven images represented, that the daily sacrifice of innocent people might be stopped.

There is not one of the readers of this work but will admit there is a force in nature that did roll back the sea from the Island of St. John till the bottom of it was left dry for miles, and then returned it with a tidal wave that destroyed ships, submerged cities, and destroyed people. But when they read in the Jewish history that the Red Sea was divided and the Israelites passed through it dry-shod, and that Pharaoh and his army, in attempting to follow were submerged, overwhelmed, and destroyed, many will say, too, why should God trouble himself about the Israelites or Pharaoh's army? Answer: That the faith of Noah in a God of infinite intelligence, love, mercy, and goodness might be carried into Africa and take the place of the fanatical belief in the minds of that people of an endless see of devouring beastly deities, and the daily human sacrifices stopped, and all the other horrid abominations practiced under it might cease.

The world is full of people, and every city has its representatives of them, who say Abraham was our father and Abraham's God is our God, and they believe in that history and every miracle recorded, because all their ancestors before them and they themselves at the present time are celebrating those providential occurrences yearly on the day that they transpired, as we the people of the United States keep the declaration of independence in mind by celebrating the day on which it was signed, and devote that day especially to speech-making and talking over the terrible struggle for independence.

Under the reign of rulers who believed in the above-described devouring beastly deities, the several petty kingdoms kept up a perpetual strife for empire, under which civilization could make no progress, and the most terrible devastation, murder, and robbery were perpetrated, down to the time that Babylon became the seat of human government, and every nation among men was represented there by their best statesmen and philosophers during the reign of Nebuchadnezzar, Cyrus, Darius, and many others, and a general mingling of Asiatic, African, and European ideas took place, and the sharp corners and rough edges of all religious faith, political systems, and philosophic theories were rubbed off and toned down to a more humane condition, and poetry and literature rose to its very highest pinnacle.

Under this stimulus the greatest dramatic work extant (the Book of Job) was produced, Homer wrote his Iliad and other gems of poetry, Confucius his moral maxims, and Plato dressed up that pagan see of deity in a much more becoming garment, and even seemed to have some idea of a god quite different from a devouring beast; still he clung to the old theory of a patern in nature for everything, striving to manifest itself, which even he could not separate from the theory that every species of animal culminated in a god, whose attributes were represented by the species of animal of which it was the type and ruling deity.

So the daily propitiatory sacrifice of innocent people went on till the advent of Christ, and his sublime doctrines and holy example were promulgated throughout Africa and Europe, by which their granite images of a beastly deity were crumbled to dust, and the daily propitiatory sacrifice of innocent men, women, and children ceased, and the whole ruling pagan dynasties who worshiped them and committed those terrible abominations of barbarism fell down before that name, as Dagon had fallen down before the ark, and their pagan heads of barbarous rule were broken off.

Constantine was the first great potentate who prostrated himself before it, and all the others followed in rapid succession, so that now for over a thousand years the world has been governed by dynasties who believe in a personal, omnipotent God, of infinite wisdom, mercy, and goodness, who created all things, and directs the course of nature by his irresistible fiat; and under its humanizing influence civilization has made greater progress in the last thousand years than it did under five thousand years' reign of pagan superstition and petty kings.

But as it was written that after the thousand years were accomplished Satan should be loosed for a season: that is, that the old idea of an endless see of a serpentile, devouring, beastly deity, which had wriggled itself up from profound ignorance, to a grade of blank don't-know-nothingism, to the intelligence of myriads of species of devouring beasts, demanding of kings the sacrifice of their innocent children and dearest friends as the price of their crowns and right to rule, was to be revived for a season and many deceived by it.

Hence we find the skeleton of it, under a patchwork of divers-colored garments, in the above paragraphs from a poem, Darwin's origin of species, and in sundry newspaper correspondence, which, when compared with the eloquent garments which Plato clothed it in, really appear quite shabby.

Neither Darwin, nor any other of the modern votaries of that old pagan beastly deity have yet, and we assert here without fear of truthful contradiction that they cannot, produce the least shadow of evidence, obtained by scientific test, to prove that, within the pale of history or the memory of man, any species of animal has ever branched off into, or thrown out, a shoot which has produced a new species, or has ever rendered any aid in the production of one. But, on the contrary, every step which scientific investigation has made in the amalgamating and crossing of races has confirmed the fact, that the animal force of each species is as constant to its type as electricity is in its chemical effect, gravity in its sphere, crystallization in its form, or the polar attraction in guiding the mariner's needle, and in all of its propagation has repeated itself in the specific form of the first pair or primitive type with the same persistence that the crystallizing force of the mineral salt forms a cube, or that of carbon forms the diamond with its peculiar angle and cleavage. The lion of to-day is exactly like the first lion that man ever saw, and wherever a cross has been made, the offspring invariably resembles its sire.

By breeding a male Angora goat with a flock of Spanish or common coarse-haired female goats, and again with the hybrids, and again with the quadroons, and so on up to the seventh generation, the animal force of the male Angora will have worked the blood and form up to its primitive specific standard, and the mohair of the seventh generating cross will be equal to the original type. So, on the other hand, breed into a flock of female Angoras with a male coarse-haired Spanish goat down to the seventh generation, and you have a coarse-haired worthless goat, a perfect model of the typical pair of worthless-haired goats, worked down by the animal force, of the primitive species.

And this is the history of the result of every cross that has been made, where the hybrid will breed at all. In a cross between a male donkey and a female horse, the product of the first cross is a great awkward braying donkey, and that is the end of it. A cross between a white man and negress, in the same manner, will result in the same way: the animal force of the man will bring the cross up to the Caucasian standard in the seventh generation. So also, on the other hand, a cross between an African negro and white woman: the animal force of the negro will work the progeny of the cross down to its original African type at the seventh generation. And so also if a woman would breed from a male monkey, and the progeny would breed in with that male monkey in the same way down to the seventh generation, the product would be a pure model of the specific monkey type.

Never since the memory of man has a cross of species stopped at any point between the two in the form of a new type that could repeat itself: in every instance it runs clean back to one or the other.

With all this evidence before them, and not a shadow of experimental evidence or proof obtained by scientific test in favor of their theory, these dogmatic revivers of an old pagan heresy stand up and affirm that the soul of man has torn itself off from a grade at least as low down as a monkey, and the poet takes it from a vegetable plant, and has made quite a respectable cannibal god of itself, and even prints it in books and journals, which find purchasers and readers and some believers, as it was predicted there would be in the last days of human sovereignty.

Except those modern votaries of a pagan deity can produce some evidence that mankind will be benefited by abandoning Christianity and adopting the pagan religion, with its inevitable king to represent each specific beastly god and the daily appeasing sacrifice and children-consuming fires, they will hardly expect to overthrow Christianity and re-establish it: the first necessary step in which is the destruction of all telegraph lines, railroads, and steamships; the next, the walling in of all cities, and the construction of castles for those specific gods and the kings who rule by their authority to dwell in and secrete their plunder, and pyramids to perpetuate the names of great tyrants; third, the destruction of all constitutional rights and vested privileges of the masses; the demolition of all their comfortable dwellings, titles to land, and liberty of speech, and force them to dwell in caves, booths, and in the open fields; subject to stripes, wounds, bruises, putrifying sores, unrequited labor, and death without retribution, and robbery without restitution; for such is the photograph which history gives us of it in its best condition. Its days are ended; neither can any power on earth restore its vitality.

Darwin's theory certainly denies or ignores the fact of an intelligent personal creator, working by architectural plan, drawn by process of reasoning, and executed by force of will out of infinite resources, in a rational, intelligent manner. Consequently he must take as the foundation of his theory one of the two pagan theories: either that each species culminates in a specific, sovereign, self-existing force, which they call a god, or that one universal law pervades infinitum, so full, that no vacancy exists in the vast immensity of space, and the said force is without any intelligent head or center of action, acting with the same equal force in the same manner everywhere. To the latter, the very first test of science dispels it as the morning sun does a miasmal fog. If there is any one question which science has established, it is that each law of nature acts constantly in one direction, without variableness or shadow of turning.

Gravity always forms spheres, crystallization cubes or crystals, and the planets are all moved in definite directions, each by its own specific force, and between the species of animals there is an uncompromising hostility, an impassable gulf of specific enmity, which positively prohibits amalgamation, except through the potent influence and agency of man. A ball set in motion by any one force, moves on in that direction till it is intercepted by a counter or opposing force; and for any man to assert a theory that one single force in nature, without an intelligent head of device, or center of action, is playing this shuttlecock game of counter movements and varying evolutions, and constructing the endless variety of specific vegetable vitality of such endless variety of forms, and of animal species with such endless variety of desires and social habits and infinite change in physical structure, and absolute enmity and hostility between species, is absurdly preposterous, and was done to death something over a thousand years ago by the progress of science, stimulated and urged forward by that progressive Christianity whose kingdom was compared to a little leaven in a measure of meal, a grain of mustard seed, and many other spreading and progressive things. That theory is not only dead, but consumed to ashes in chemical laboratories, and scattered by the four winds of heaven, called ethics, science, mathematics, and Christianity, which by working together have made wonderful progress.

The other pagan theory, that each species of animal culminates in a god, or a sovereign, independent, self-existing, and specific organizing force, knocks Darwin's theory of monkey origin for man out of the theoretical arena, and leaves it without a shadow of plausibility, and explodes the whole theory.

It is amusing to see a man with sufficient intelligence and education to write a book spend his life in trying to trace his origin down through a line of Digger Indians, caudal negroes and negroes proper, bushmen, chimpanzees, orang outangs, gorillas, and baboons to a monkey ancestry! Some people may be proud of a baboon or monkey ancestry, but they are scarce.

The object of this work is to give a rational theory of the origin of species, and we request the reader to give it a fair hearing, read it attentively, and digest it thoroughly; and if it is not sustained by practical progressive science and sound common sense, as well as Christianity, reject it. We do not ask you to accept it without good evidence.

We have deemed it proper in this introduction to briefly notice a few of the most important ancient dogmas which have hitherto led to such confusion and misunderstanding of the origin of animal species. There are, however, some others that will be noticed in the following work; and we will further state here, that the investigation will take us into the upper regions of spiritual life, where the combinations of simple spiritual elements that constituted the specific propagating type in the beginning, which have existed ever since the historic records and memory of man, have been repeating themselves by propagation with the same constancy that the seed of an apple has produced its typical tree, or the sun has repeated its light.

We of course have to get up into that spiritual region by analogy, as Newton got up among the planets and investigated the laws which control them.

It is only a few centuries since the science of the movement of the heavenly bodies was no better understood than the science of intelligent movement now is; and the field of the latter is just as open for investigation, and quite as new to us and as easy of comprehension, as the former.

CHAPTER I.

FORMS AND SOCIAL HABITS OF SUNDRY TYPES AND RACES.

As the investigation of the simple elements which enter into the composition of that mainspring of animation, "and being immutably combined, constitutes the principle called soul," leads us into a field of natural philosophy hitherto but very little explored by scientific research, we give in this chapter a sort of discursory statement of sundry well-established facts, relative to the specific forms and social habits of sundry types and races, for the purpose of preparing the mind of the student to explore this new field in an earnest, orderly manner.

Wherever there is animal life there is more or less power of self-action, nerve, and sensibility, and the causes or principles which constitute the sensibility and instigate the action are as easily comprehended as those of astronomy or chemistry, and the field is as open for investigation, and quite as full of interest, to the enthusiastic explorer.

Vital science is naturally divided into two great branches: one comprehending the cause of that great variety of specific plants in the vegetable kingdom which have an organic body without sense and spontaneous motion, and which usually draw their nourishment partly through roots and fibers fixed in the earth, and partly through leafy lungs from the atmosphere; the other, all that variety in typical forms and social habits of races and individual eccentricities of the specimens of the same race in the animate kingdom. One may very properly be termed the philosophy of animal life, the other the philosophy of vegetable vitality.

Our present investigations lead us through the field of animal life; and our first work will be to examine some of the forces which control its movements, after having examined some of the species which constitute the subject under consideration.

In examining that field, we find several groups in the great chain of animal life, composed of several ascending or descending links, which have something of a similarity of form, but differ widely in their work and habits of life, and each specific link persistently propagates its own kind.

At the head of this great chain of animate nature we find a group of links descending in regular lessening grades from man to a very small monkey, classed mammalia, which walk mainly or altogether upon two legs and suckle their young at the breast. Commencing at the small end of this chain, we find a specimen of it which very much resembles a squirrel, but walks upright at its pleasure, called the tarsier, and between that and the baboon several ascending links, described in detail by Rev. J. G. Woods's Natural History, which never cross in propagating their species, any more than the striped, red, and black squirrels.

Next in the ascending scale we find the gorrilla, orang outang, chimpanzee, bushman, caudal negro, negro proper, and several different species of Indians, some of which approach about as near to the form of man as a squirrel does to a petit monkey, or an ape does to a negro.

We also find another group of domestic animals descending by regular specific links from the elephant to a rabbit, whose habits are gramniverous; another group of carniverous animals, from the lion to the dog; another from the tiger to a very small cat, all going upon four legs; then there are several groups of ants and other insects going upon six legs.

We also find several groups of centipedes, worms, and larvæ, which travel upon a great number of legs, indeed their most striking feature is legs; and there are a great number of groups in the fowl and bird species which have two legs and two wings, and walk or fly at their own pleasure; and several groups of snakes, serpents, and worms which have no legs at all, and move about by a wriggling movement, or expansion and contraction of the body.

We also find a great variety of groups in the piscatory department of the animate chain of creation descending by regular graded links from the great whale down to the lamper-eel, whose propagating element is water; and numerous groups of monads, whose habits of life are only discernible by the aid of a microscope.

For full and entertaining descriptions of the forms and habits of all the specific varieties of the entire chain of animate life the student is referred to the Rev. J. G. Woods's Natural History.

Each specific link in that immense chain of groups is a generating family, modeled by an immutable type, which has repeated itself in the same specific form by propagation, about equally paired as to males and females so long, that the memory and records of man runneth not to the time when the contrary was known to have been done; the males and females of each type always pairing together for domestic enjoyment and copulating purposes, never crossing in marriage, pairing, or amalgamating with the upper or lower consecutive links of the same group, however much the form or habits of life may resemble each other. As, for instance, horse and donkey, of their own accord (mules, mulattoes, and hybrids of every description, are produced by the instigation of man,) cows and buffalo, leopards and tigers, wolves and bears, deer and elk, geese and ducks, chickens and grouse, wasps and bees, red and black squirrels, apes and negroes, or negroes and Indians, chimney and barn-swallows. As the poet sung,

"Like loves like, and love likes love;
Eagle seeks eagle, and dove seeks dove."

And their round of social enjoyment and domestic felicity is all confined within the circle of their own propagating link. In fact, the absorbing object of all seems to be propagation of their species—to feed, protect, and rear their young. The parents will fight and risk their lives in every conceivable manner, to the very door of death, to shield and protect the young till they are large enough to take care of themselves, for which paternal care the young make little or no return of grateful care for their aged parents, but go on to the production and care of a brood of their own. So it seems that the main object of this terrestrial life, with all species, is propagation.

However much any two propagating types may resemble each other, a close examination will show a marked difference in form of body, texture of flesh, shape of bone, flexibility of muscle, and tension of nerve, which somewhat changes the tone of its enjoyment, power of action, and quality of its food.

The wasp feeds upon meat; the bee upon the nectar of flowers; cats upon flesh; rabbits upon vegetable substances; sheep upon vegetables; while hogs are both gramniverous and carniverous. Thus all the specimens of each specific link voluntarily seek the food which affords the most nourishment to their own body.

Concerning this difference in the shape of bone and component quality of the flesh and blood of the various species of animals, the Rev. J. G. Woods says:

"Thus it will be seen how easily the observer can, in a minute fragment of bone, though hardly larger than a midge's wing, read the class of animal of whose framework it once formed a part, as decisively as if the former owner were present to claim his property. The life-character is enshrined and written upon every sanguine disc that rolls through the veins, is manifested in every fiber and nervelet that gives energy and force to the breathing body, and is stereotyped upon each bony atom that forms part of its skeleton framework. Whoever reads those hieroglyphics rightly is truly a poet; for to him the valley of dry bones becomes a vision of death passed away, and a precursor of a resurrection and life to come. * * *

"Not only is the past history of each living creature written in every particle of which its material frame is constructed, but the past records of the universe to which it belongs, and a prediction of its future. God can make no one thing that is not immortal in its teaching, if we would be so taught; if not, the fault is with the people, not with the Teacher.

"He writes his own living words on all the works of his hand; he spreads this ample book before us, always ready to teach, if we will only learn. We walk in the midst of miracles with closed eyes and stopped ears, dazzled and bewildered with the light, fearful and distrustful of the word,

"It is not enough to accumulate facts as misers gather coin, and then to put them away on our book-shelves, guarded by the bars and bolts of technical phraseology; as the facts must be circulated and given to the public for their use. It is no matter of wonder that the generality of readers recoil from works on the natural sciences, and look upon them as mere collections of tedious names, irksome to read, unmanageable of utterance, and impossible to remember. Our scientific libraries are filled with facts, dead, hard, dry, and material as fossil bones that fill covered and sealed libraries of the past.

"But true science will breathe life into the dead mass, and fill the study with poetry and interest."

Each repeating link in the chain of animal life is made up of a multitude of individuals, each of which is a self-acting institution, having a little circle of its own enjoyment, all concentrated in producing as numerous a family as possible of its own: so that each specific link is a chain of family links, and each family link a chain of individual links, all concentrated and held together by a copulating sympathy and paternal attachment, which maintains an almost impassable gulf between them and other species in all the enjoyments of the social circle. And every movement of each individual specimen is the result of a will, produced by a belief that the movement will be for its benefit or enjoyment; and in their social intercourse will instruct, flatter, and tease each other in various moods: even get mad and fight like gladiators, till one or the other is vanquished.

The members of one repeating link frequently feed upon the members of another species, but never upon their own—as men, dogs, cats, lions, tigers, wolves, eagles, hawks, owls.

There are three attributes or necessary properties inherent in each repeating link, that are absolute barriers to feeding upon their own kind:

First. A copulating affinity, which irresistibly draws the male and female of the same species into a family circle for the increase of their own kind.

Second. A paternal affection, which gives the most tender care for the young till they are capable of providing for themselves, and a filial regard for the old and decrepid, and a desire to make a common cause in providing for the crippled and infirm.

Third. A fraternal sympathy, that inspires a fellow feeling which is ready to battle for every member of the species against all outside barbarians of every species, and draws them to the common center of a social circle, both for social enjoyment and mutual protection, and thus forms nations, tribes, flocks, and swarms for self-protection against other specific links, impressed with the idea that in their aggregate capacity they are much better able to cope with their enemies than they are in their individual capacity. Consequently, if cats fed upon cats, wolves upon wolves, lions upon lions, hawks upon hawks, and eagles upon eagles, apes upon apes and hogs upon hogs, the strong would devour the weak, the old the young, and the race would be exterminated in the first generation; for the young cleave to the old and the weak to the strong for protection.

We have made this discursory introduction in order to shake, if possible, some of the old prejudices of the reader, and prepare his mind for a careful perusal of this work and a thorough investigation of this subject.

CHAPTER II.

THE SOUL—ITS STRUCTURE AND MOVEMENT.

Vital science, like that of natural science, is divided into two great branches: the one comprehending all perceptible movement which is the result of volition; the other including all those natural changes which produce a tree from a pit, nut, or seed, and all manner of vegetation reproduced by seeds after its kind by germination. The first is called the science of sensibility, including the primary principles controlling volition, which determines the similarity of form, actions, enjoyments, and habits of life of each copulating species of living creatures in the animal kingdom, from man to a monad.

The second is the science of vegetable vitality, comprehending the primary principle which gives vitality to vegetation, including the causes of the almost endless variety of form in trees, shrubs, vines, and grasses, beauty of flower and blossom, quality of fruit, form and vitality of seed, and secures an exact likeness of the original type in the product of each seed, germ, or shoot.

The science of sensibility is of the greatest importance to mankind, since by its investigations a clear line of discrimination between the propagating species of animals is discerned, and the disastrous results of amalgamating with the lower species clearly demonstrated.

The science of sensibility is also divided into two great branches: the one comprehending the spiritual and celestial combinations of principles which give vitality and volition to the animal kingdom; the other the terrestrial forms and movements produced thereby. The former treats of the nature and organic structure of the soul; the latter of the material form and movement or work of the soul.

That the vital principle called soul, which gives animation to the animal kingdom, is a combination of celestial substances and spiritual forces, is demonstrated by the peculiar mode of mental enjoyment and kind of food adapted by each species. In confirmation of which Paul speaks of the dividing asunder of joint and marrow, soul and spirit; and Moses, in giving a history of the creation of the soul of man, says: "And the Lord God formed man of the dust of the ground, and breathed into his nostrils the breath of life; and man became a living soul."

The body was not a soul, nor the breath of life a soul, but the combination of the two constitute a living soul, with a substantial body, having form and certain powers of voluntary movement, both visible and effective: a self-moving institution, an animal microcosm.

Thus the constituent elements of a soul are taken

partly from inert material substance, and partly from active spiritual essence. The former embraces all those inert gases which, under the influence of the laws of nature, enter into the combination of all material bodies, and constitute the warp and woof of every plant, rock, grain of sand, and particle of dust, which can be perceived by the finite senses of seeing, hearing, smelling, tasting, or feeling, as well as all vegetable forms which grow out of the earth, and every animal carcass that feeds upon it; and has neither sensibility, affinity, power of motion, nor combination, and left to itself would eternally remain an inanimate element, useless and harmless, simply helping to prevent a vacuum in the immensity of space, without any possibility of ever changing its proportions.

The latter comprehends all the spiritual elements of volition which enter into the composition of all the specific degrees of sensibility and reasoning capacity which constitute the graded links in the great chain of animal life, and is as constitutionally active as matter is inert, and each simple element can no more cease its peculiar action than matter can begin to act: a spiritual spark can no more stop its spontaneous motion than matter can begin to move.

The infinite variety of results produced by changing the combinations made from the few simple elements of matter is naturally divided into four great branches.

The first comprehends all that variety in form and material found in crystals, rocks, earths, metals, and fluids in the mineral kingdom.

The second, all that variety in material and form of plants, beauty of flower, and fragrance of blossom, flavor of fruit, and pungency of seed, in the vegetable kingdom.

The third, all that variety in the form of animal bodies, and quality of material of which they are composed, in the animal kingdom.

Fourth, the same variety of form and material in the celestial bodies of animals, which constitute the leverage and working machinery of the soul.

For example, the same gases combined in one way produce water, and in another proportion atmospheric air. A certain other proportionate combination of simple gases constitutes the diamond, and another quartz crystal; another rocks; another earths; another metals; another plants, flowers, fruits and seeds; another the bones, flesh, muscles, and nerves of the terrestrial bodies of animals; another and most important combination produces a corresponding celestial body, constituting the leverage and machinery of the soul, which though as invisible as the gases which form the diamond, air, or water to our terrestrial senses, still is as immutable in specific construction and perfect in bodily form as the terrestrial body in which it commenced its existence. Neither have the gases of which it is composed any more power to organize that body than iron, copper, trees, water, and caloric have to organize themselves into a steamship or railroad and rolling stock for the benefit of commerce.

The whole work of combining gases, organizing spheres, and constructing bodies, is the work of organic forces. Even the celestial body of the soul is formed by the force of animal growth which inhabits it by the prior right of construction, and is always busy keeping it in repair; while the spirit of sensibility, in consideration of a life lease, endows it with volition, sensibility, reason, and enjoyment.

The simple elements of spirit, in their various combinations, are the mainspring of all volition, maker and executor of all laws, and prime cause of all sensibility: an intelligent investigation of which is the object of this work. A simple element of force (which may very properly be termed mites of force) is an active essence, endowed with a certain power of self-action and a specific form, which can only demonstrate itself through the agency of a substantial body, over which it has entire control, and always constructs from the simple gases, by its own constructive power, in the precise form of body of which itself is the pattern.

Every material body into which the simple elements of matter have ever been formed has a spiritual pattern, endowed with a self-constructing force; and, incomprehensible as the variety is, they are all produced by a change in the combination of a few simple mites of forces.

The forces which thus organize the simple elements of matter into substantial bodies and endow them with attraction, vitality, and sensibility, are naturally divided into four spheres of action.

The first comprehends all that variety of combinations from the simple elements of action which organize all synthetically constituted bodies, whether of earthy minerals or planetary spheres, and controls their movements in the planetary system, and is denominated the force of natural law.

Second. All those elementary combinations which constitute the forces of electricity, earthquakes, tornadoes, nitro-glycerine, &c., are denominated explosive forces.

Third. All of those combinations from the simple elements of force which constitute that endless variety of the specific forces of growth in the vegetable kingdom are denominated vital forces of vegetable growth.

Fourth. All of those combinations from the simple elements of force that enter into the combination of the forces which construct the specific bodies in the graded links of the universal chain of animal life, with supplies furnished to every part of the body through a network of veins and arteries, and has its seat in the heart, where they are all concentrated, and perform the whole work of constructing the body, repairing all damages, and making an entire change in the whole material once in seven years, but has no more power to endow that body which it has made with sensibility or a reasoning capacity, volition, or motion, than the force of growth which has formed a plant has to endow the body of the plant which it has constructed with the same attributes. Spirit alone can produce sensation, sensibility, reasoning capacity, and enjoyment on the brain and nervous system after it is formed: this force is called the vital force of animal growth.

Thus we have a clear line of distinction drawn between explosive forces and forces of natural law, vital forces of vegetable growth and the vital forces of animal growth, and between all manner of forces which manifest no intelligence and the intelligent spiritual forces which fill the universe with intelligence, joy, and gladness.

For instance, the work of explosive force is to destroy organic bodies; the work of constructive force is to combine gases and work them into the form of which itself is the pattern. Thus, we know that gravity is the pattern of a sphere, because it organizes everything in that form, and that all crystallizing forces are a pattern of a crystal, with a certain angle and cleavage, because they organize all their bodily representatives in that form; and we also know that all the vital forces of vegetable growth are a pattern of plants, for they construct all their bodily representatives in that form; and we also know that all the vital forces of animal growth are a pattern of some animal carcass, for they construct all their bodily representatives in that form; and we also know that the attribute of spirit is intelligence, for wherever there is volition and power of locomotion there is a spirit and some degree of intelligence, and the degree of intelligence as well as the form of the body depend entirely on the equivalent proportions of simple elements combined in the soul.

CHAPTER III.

ELEMENTARY PRINCIPLE OF SOUL.

A simple element of spirit is an intelligent essence, having a certain power of action, but is incapable of demonstrating either without the aid of a body toned with a nervous system. But as all of its powers of action are for the purpose of working out some degree of intelligence and nervous sensation upon a brain in the concentrated terminus of a nervous system, having no power for the combination of gases or the construction of a body of any kind, left to itself it is as powerless and harmless as a simple element of matter, but when combined, or rather associated, with a congenial force of animal growth, which has power to form a body with a bone and nervous system peculiarly adapted to its power of sensible action, it commences immediately to demonstrate its degree of intelligence: first, in the form of the body which it has planned and superintended the construction of; second, by its intuitive work and mode of life.

Thus we discover that the simplest, smallest possible soul, which is supposed to be the least of the monad species, is composed of three elementary principles: First, an intelligent spirit; second, a force of animal growth; third, a substantial body, composed of celestial gases, all of which are as invisible to our terrestrial senses as our terrestrial bodies are to a blind man; but nevertheless as immutable in bodily form, and perfect in specific intelligence and capacity of enjoyment, as it was in the terrestrial body in which it commenced its existence.

As the soul of the smallest specimen of a monad is supposed to be composed of a very small number of simple elements of spirit, and the smallest possible number of mites of simple force of animal growth, and a very small number of simple gases, then it is evident there must be several simple elements of spirit, as well as of force and matter, out of which to combine equivalent fractional parts in just the proper quantity to produce the intelligence with which man is endowed.

As the combining of the simple molecule elements of matter into substantial bodies of various magnitude and hardness, and the simple mite elements of force into patterns of an infinite variety of forms, and the simple intellectual elements into a capacity of intelligence to fill the circle of useful inventions which man has already accomplished, is entirely out of the power of man, therefore the only resource of science to discover the number of the simple elements of each is by analysis or analogy.

By the aid of chemistry, any ponderable body, even the diamond, can be decomposed and converted into its elementary gases, and something of a test made of the part they act in giving density and hardness to bodies; and thus chemists have been able to name several which are known to be simple distinct elements: among which are oxygen, hydrogen, nitrogen, and carbon, which are known to be simple elements, and are found in various proportions of combination in almost all bodies; chlorine, iodine, bromine, florine, and phosphorus are also believed to be simple substances:—altogether seven simple gaseous elements, out of which to combine and constitute all the myriads of fluids, solids, vegetable and animal bodies, which are found in this earthly sphere. And the greatest number of simple substances, including metals, which chemistry cannot prove to be compound, is sixty-one, and the probability is that when these are reduced to their original simple ethereal elements, they will be found to resolve themselves into seven simple specific gases, out of which the earth and all that pertains to it have been formed, and they are all as ethereally invisible and incomprehensible to our terrestrial senses in their simple elements as the constructive forces or the spirit of life itself; but in their combined, fluid, crystallized, and solidified condition, we see, hear, taste, smell, and feel them, and discern their qualities, and can analyze and trace them to their simple elements, and in so doing we find that in their simple elements there is no attractive affinity between them;

but their whole tendency is repulsive, and to fly off as far from each other as possible. However little there is in a bottle, it is always full: that is, they pervade the whole of it, and never settle together in any one part, like water or other substances.

Thus it is positively demonstrated that they would never constitute any body by affinity, nor was this ever done except by a controlling force of combining law, or vegetable or animal growth, which, if we could analyze and reduce to their simple elements, we would find resolved themselves into about the same number which it takes of days to make a week; and as these simple elements of force in their elementary conditions have no more affinity for each other than the elements of matter have in their state of gas, it necessarily follows that there must have been a combination of the simple elements of force in order to produce a force of any specific effect.

In their elementary, uncombined state, each mite is dancing its own peculiar jig; and in very fine dust, under a microscope, may be seen to toss it about in a very lively manner, not tending in the least to make a solid body of it, and in that state are of no more use as organic forces than twenty-six letters arranged in an alphabet are for communicating ideas as representatives of words in language. But when combined by an Infinite Intelligence in equivalent proportions for an earthquake force, let those who have felt one-half of the American continent vibrating, and seen cities and castles destroyed, judge of its effect; and when those seven simple elements of dancing molecules are variously combined by the same great Architect into organic forces, we see the use of them, not only in the wonderful things pertaining to this earth, but in the brilliance of the starry heavens. And when the same useless dancing simple forces are combined into sundry laws of planetary motion by the same Infinite Designer, with what wonderful order the planets march into position, and course through their orbits without variableness or shadow of turning.

When we look into the vegetable kingdom we see an almost infinite variety of typical constructing forces, the pattern or type for every one of which was constituted by equivalent changes in the combination of those same useless dancing mites, as existing in their seven specific elementary spheres of action. What symmetry of plants; what beauty of foliage and flowers; what fragrance of blossoms and delicious flavor of fruit; what delightful landscapes and ornamental shrubbery does the botanist discover in this field of natural science, this park and pleasure-ground of the Infinite Creator of all things. But through the whole of it, from the simple molecule element of matter and the mite element of force up to this point, there is not a single ray of intelligence, not a single nervous sensation, idea, or volition. In the hands of the Creator it is a mere pastime, and is to him simply the work of his own hand,—his steamship, railroad, plantation, park, book of prose, poetry, castle, library, palace, throne, and crown of gems, and garments studded with brilliants,—but no joy or intelligence outside of himself: and if, as some have vainly imagined, there was none even in himself, then what was the use of it? Why combine those dancing mites into such exquisite patterns of constructing forces, and those ethereal elements of matter into such admirable forms, gems, and landscapes, which are void of appreciation, and rejoice not in their existence? Neither is there one iota of intelligent spirit in the composition of them.

Spiritual intelligence is the organizer of forces, but not the power which drives them. The simple elements of intelligent spirit which is the mainspring of all volition, the endowment of all intelligence, the instigator of all joy and sorrow, pain and pleasure, happiness and despair, are as constitutionally different from the simple elements of the organic and constructive forces as they are from the simple elements of matter, and if John the Revealer's vision is true, there are seven of them, which he says he saw burning in seven golden candlesticks before the throne of God; and the combinations from them present about the same number of species and variety of form in the animal kingdom that there is in the vegetable kingdom produced by the force of vegetable growth, and in the mineral kingdom by the forces of solidification and crystallization: therefore we judge that the simple elements of force in each kingdom are about equal, as the effect of the combinations is about equal in number and specific variety in the three spheres or kingdoms into which the organic and constructive forces are divided; and all specific change in the combination, or rather all specific effect to be produced by a change in the combination, of the simple elements in the three kingdoms was exhausted ere the creation ceased. When God commences a work he finishes it, and leaves nothing undone that can be done in that field. Man, who was the last combination, exhausted the changes, so that no new vegetable substance, species of plant, or animal species, has been discovered within the pale of history, or since the time to which his memory runs; neither has any species of animal, vegetable, or crystal changed its typical form; each, in repeating itself, has been as constant to the specific pattern of its type as the earth has to its annual and diurnal movements, and the moon to its lunar circuit.

In proof of our hypothesis that there are seven simple elements of intelligent spirit, seven simple elements of organic force, and seven simple elements of substance, out of which all this wonderful creation of animal intelligence, vegetable vitality, and substantial bodies have been constructed, grown, and endowed with mobility, sensibility, and reason, we have traced matter to its seven simple elements; and John the Revealer testifies to the seven simple elements of spirit; and as the effect of change in the combination of constructive forces agrees in number with them, then the proof is positive as analogy can make it, and is further sustained by the number of simple sounds and simple colors and the number of specific effects produced by a change in their combination: which will be more fully considered in another chapter, in explaining the mode of making the said combinations.

CHAPTER IV.

EMINENT RELATIONS BETWEEN SUBSTANCE AND FORCE.

The reader who has carefully perused this work up to the close of the last chapter, has discovered that the animal which Moses said God made out of the dust of the earth, and breathed into his nostrils the breath of life, and he became a living soul, has concentrated in his terrestrial person equivalent extracts from all the simple elements of force and substance which entered into the composition of the world he inhabits, and is in the fullest sense of the term a miniature world, a small representative of the great world, a complete microcosm, partaking to some extent of the nature of every element of which the great world is composed, .*i. e.*, with the addition of the spiritual element.

The forces which form his person and endow it with volition and intellectual powers are composed, in part, from the elements of intelligent spirit, and partly from the elements of organic force, by a combination, or rather a harmonious union,—a sort of copartnership between an equivalent force of animal growth and a congenial force of intelligent spirit; and in order that their work may be harmoniously conducted, and no interference between the bodily repairs and intellectual pursuits, the force of growth has its seat in the heart, and the force of intellect in the brain, the two being individually combined, each from its own simple elements, in separate proportions, which is just equivalent to a harmonious union of all their operations. The force of growth forms and repairs the body after the exact model drawn by the spirit; that is, the pattern of the specific type in each is exactly alike, and the tone of the nervous system is toned in perfect harmony with the spiritual intellect. Thus they commence the work of life together in the ovoid, and thus they work out the destiny of life together, from the period of conception through the endless rounds of eternity, the animal force building, repairing, and toning up the system, operating from its seat of vitality in the heart, and the intellectual spirit quickening it with sensational enjoyment, power of volition, and a perpetually beating around the little circle of reasoning capacity and inventive work, which the equivalent combinations from two or more of the simple elements of spirit have endowed it with. Thus it is evident that no soul could exist without this concurrence in form and co-operation of work between an equivalent force of animal growth and intellectual spirit, both of which are so ethereal that they could never be brought in co-operation, nor able to demonstrate their form, work, or intelligence without the aid of a substantial body, adapted to an endowment of volition and intelligence.

That body in which man made his first bow to the world was also composed of elements from two distinct fields of substances: one we call celestial substance; the other terrestrial matter. The former is of too ethereal a nature to be brought within the purview of our terrestrial senses. The retina of a terrestrial eye is too gross to receive an impression of it or to reflect it; the drum of the terrestrial ear is too thick to vibrate at the sound of its voice; the terrestrial nostril has not sufficient sagacity to detect its fragrance; the terrestrial palate cannot discern its flavor; neither can the terrestrial nerve detect its touch. Consequently the chemical laboratory has no power over it, neither can any analytical test be made of it: therefore, when once formed into a body, nothing can decompose it but the omnipotent force which formed it. Terrestrial matter alone is subject to decomposition when once formed into a substantial body; but when we are released from the terrestrial element of our bodies, that celestial body is as visible to our celestial eyes, and its voice as distinctly heard by our celestial ears, and a celestial person will be as readily comprehended by our celestial senses, as a terrestrial body is to our terrestrial senses: but no eye will ever see an intellectual spirit or force of growth, nor the form of an organic law.

In proving this hypothesis, it is not necessary to

call as witnesses the seers of the olden time, nor Paul, nor John, nor Swedenborg, nor Andrew Jackson Davis, nor even modern clairvoyants. All that any person has to do is to review his own experience.

Who is there that has not dreamed and remembered and told his dreams? When the intellectual spirit has ceased its agitation of the nervous system, and given it over into the hands of the force of animal growth to repair the damages of the day's contests, and it is stretched out, a vital emblem of a corpse, in an insensible condition, while the repairs are being made, who is there that has not stepped out of his terrestrial habitation, and not only seen but conversed with his departed friends, and been warned of coming dangers or impressed with coming joys?

Thus every person is convinced by his own experience that within his terrestrial body, and pervading every part of it, there is a celestial body, over which the explosive and decomposing forces of terrestrial bodies have no power, and that there must be the same number of simple celestial elements out of which the combinations are made; for the effects are equal, and the variety of combinations uniform. Without this celestial body there could be no soul; and that body, once formed, is as imperishable in its constituted form as the planetary system or the throne of Jehovah, and nothing but his fiat can ever separate it from the spiritual and constructive forces that formed it. Hence the unquestionable immortality of the soul. Being constituted of celestial substance, force, and spirit, it can neither be decomposed nor cease to exist.

In this refined celestial state the soul has no power of propagation, and lives exclusively for its own enjoyment and the set in which it moves and associating companions. A terrestrial body is an indispensable requisite to propagation.

Concerning the terrestrial part of the animal body there needs but little explanation. We are encumbered with it; we are tortured by its injury, pained by its repairs, grieved by its short-comings, and anon overwhelmed with joy by its pleasurable excitements. Chemists have analyzed it, and found it to be combined of the same simple elements of which the earth is composed, and found it to be a combined extract from nearly all the terrestrial gases, constantly changing old for new matter, and requiring nightly repairs of the breaches and bruises and losses it has sustained in the exciting labor of the day; and when the soul gets tired of this doctoring and patching and repairing, and bids it a last farewell, and passes away to the celestial sphere, it is readily decomposed by the surrounding chemical influences, and returns to its elementary condition, except the teeth and bones, which stand the test of nature's laboratory much longer. It is merely a breeding tenement, and bears a similar relation to the soul that a larva does to a miller and a chrysalis to a butterfly: a copulating bed, in which the soul repeats itself and increases the number of its species: a terrestrial model, in which the force of growth casts a brood of new specimens in the specific form of which itself is the pattern, and it is of so perishable a nature, that twenty-five years is about the average length of time that the soul of man can manage to patch it up as a tenement for that purpose. But without its assistance no new souls could be propagated, no increase in the family of a species could be produced. And even the simple elements of this terrestrial body are of too aeriform a nature to be seen with the terrestrial eye, and are called material gases; but when properly combined, become solids, crystals, diamonds, bones, muscles, nerves, flesh, and blood, in which we begin our existence: but when at large, their relations to each other are repellant, so that an overpowering force is required to make any substantial combination of them.

The simple elements of celestial substance bear the same repellant relations to each other, and can only be forced into a combination through the aid of material fluids and bodies, and are called celestial gases or elements of substance.

The simple elements of organic force bear the same repellant relation to each other, and can only be forced into a combination for a useful purpose by an overpowering force as potent as that which condenses steam for a propelling power, without which each mite would dance its own little war hornpipe throughout the endless rounds of eternity, of no more use than floating steam.

That those mites pervade the atmosphere as plentifully as the material gases is easily demonstrated. Pulverize any substance, even flint glass, to an impalpable powder, so that a particle of it is reduced in weight to within the power of a mite to move it, and place a small quantity of these impalpable molecules under a microscope, and you will see them tossed about in a very lively manner by those mite forces, not in the least tending to a massive combination. Thus those elements of organic force float through the atmosphere, mingling with the material gases in the same atmosphere, pushing and thrusting and tossing them about in the most useless frolicsome manner, as steam when released from the pipe of an engine makes a sudden commotion, but to no useful purpose. Its mechanical service was rendered in a condensing cylinder. So with the mites of organic force: they must be combined by as potent a force to become a constructing pattern of any thing.

The simple elements of intellectual spirit also bear the same repellant relation to each other. Each atom, content with a bare consciousness of its own existence, would loaf through the immensity of space forever, except it was combined by an irresistible force, and in that passive state of existence they are as useless for any intelligent purpose as iron in the mines is for mechanical implements. The individual specimens of the simple elements of intelligent spirit we call atoms. That these spiritual atoms pervade the atmosphere in about equal proportion with forcible mites, celestial gases, and terrestrial molecules, is clearly demonstrated by the daily propagations of new souls which they endow with brilliant intellects, and the mighty throng of new-born souls in the animal kingdom, which are constantly marching on the stage of action through the door of copulation, and up to their celestial home through the dark curtain of death; and each conception makes a draft on those floating spiritual atoms, forcible mites, celestial gases, and terrestrial molecules. The lungs of the male collects them in the process of breathing, and forces them into the veinous circulation, from which the testicles secrete them, and combine them into the specific form of the pattern manifested in the first soul of the species, and there it becomes an independent self-acting miniature soul, from which it is discharged into the ovarium of the female through the act of sexual intercourse, and finds its way to the ovoid which she has prepared for its sustenance, in which it commences the work of life, where it is nourished and nursed up to a state of ability to provide for itself. Thus the soul of man is formed, and thus the soul of a monkey, cat, fowl, fish, and insect, all from the same aerial elements of animal life; and the difference in form of body, muscular strength, intelligence, and reasoning capacity is owing to a difference in the equivalent proportions used in combining and constructing the pattern in the form of the primitive soul of the species.

The form of the pattern fixed in the first pair of a new creation of animals determines the specific form of body, strength of muscle, intelligence, and work or habit of life of the species to the end of their ability to propagate, if such a time ever comes. Neither can they change it any more than a diamond can change its brilliancy, a crystal its angle and cleavage, water its refreshing effect, the atmospheric air its nourishing qualities, or electricity its explosive force. And there are very strong proofs of the correctness of the above hypothesis; for there is about the same difference between the intelligence of a man and a monkey that there is between the density and hardness of a diamond and that of water.

As the souls of animals gather their typical spirit, force, celestial substance, and terrestrial matter from a common atmosphere, and work it into their own specific pattern in texture of flesh, strength of muscle, tension of nerve, sensitiveness of brain, strength of bone, sensibility, and reasoning capacity, in persevering constancy to their primitive pattern through all their propagating generations, so do the vegetable forces spread out their roots and fibrils through the same soil, interlacing each other in all directions, and they shoot out their branches, interlocking each other in a tangled net-work of limbs and twigs, and float their leafy lungs in the same atmosphere, gathering in their vital forces and inert matter from the same elements out of which each specific plant repeats itself in the pattern of its primitive type, each reproducing the specific form of plant, fragrance of flower, quality of fruit, and pungency of seed—as it was said in the olden time, "every tree bearing fruit after its kind."

Out of the same earthy solutions and atmospheric vapor the oak produces acorns, the apple-tree apples, cherry-trees cherries: growing grain, corn, wheat, rye, oats, peas, beans, and barley from the same soil, with their fibrils interlacing each other; and crowded in among them may be seen the thistle producing thorns, and the fig-tree figs. Thus in constancy of species the vegetable and animal kingdoms are equal, and both draw their simple elements from the same sources.

CHAPTER V.

DEPARTMENT OF SOUND.

Another department of the universe is represented by sound, the property of which is signification, and next to light is the greatest medium through which souls acquire knowledge.

Expert linguists in the science of sound have demonstrated the fact, that in the department of sound there are just seven simple elements, and have made that the axiom on which language and music are based. Every sound which a soul takes cognizance of brings with it a sense of significancy, and if often repeated, the cause is searched out and the signification demonstrated and remembered; and among men it is not only remembered, but recorded in books for the instruction of the coming generations, which is termed written language. Man is the only animal which has a written language, but every other species of animal has a vocal language, by which two or more may converse, to a greater or less extent, exchanging their limited number of ideas with the same satisfaction that men do their myriads.

For the purposes of literature linguists have invented certain characters as representatives of sound: some languages have more and some less.

The English language has twenty-six characters, seven of which represent simple sounds, and nineteen represent derivative sounds.

The twenty-six characters representing sound in the English language, when arranged in the alphabet, are of no more use for purposes of language than atoms of spirit, mites of force, and molecules of matter; in their simple uncombined state each has a sound which ends in the character itself, and that is all the significance there is in it. But when combined into syllables their significance increases, and when syllables are formed into words they have a different import. When words are constructed into sentences their purport is more significant, and sentences combined into history, essays, scientific instruction, mathematical problems, arguments, addresses, speeches, harangues, orations, and poetry, become the axle, lever, and machinery by which governments are regulated, religious worship conducted, commerce carried on, and all society organized and disbanded; political factions begin and end, kings are made and unmade, and kingdoms and empires rise and fall. They are to the intellectual spirit of the soul what the simple elements of substance are to the body, and the same simple sounds are as repellant to each other as are the simple elements of either department of the universe; for no syllable can be formed of two or more vowel sounds and give each vowel its full simple sound.

In the formation of syllables, one or more characters representing derivative sounds must intervene between two vowels; hence the necessity for nineteen characters representing derivative sounds, for seven representing simple sounds; and like all other simple elements, they have to be combined and recombined into equivalent structures to produce any effect; in fact, the only effect of any simple element in the universe begins and ends in itself; at least in all the foregoing departments the simple elements repel each other, and all useful combinations necessarily must be made by a force outside of and independent of them, and necessarily for some definite purpose, and every purpose must be preceded by a design, and every design is the effect of some process of reasoning which soul alone has capacity for, and an intellectual spirit is the mainspring and sole conductor of the whole process of reasoning by which a design can be formed or a purpose marked out; therefore, whenever a sound is heard, if traced back to its first cause, it will be found to be the effect of a design which was worked out by a sensible soul; and the combination of every body or substance visible to our terrestrial senses is the work of an independent intelligent soul.

The souls of animals, even man himself, have no synthetic influence over the simple element of the first four departments already considered, except to construct their own bodies out of them in the same insensible manner that a vegetable performs similar work. The forces for organizing them can only be constituted by an infinite intelligence. Neither has man power to decompose anything but terrestrial matter.

The chemist, with all his explosive forces, intensified caloric, and corrosive acids, can touch nothing but matter: acids, fire, and electricity have no more analytical effect on the substance of which the celestial body of a soul is composed than nitric acid has on gold; and even over the simple elements of that he has no power of organism, he can only decompose. In vain the alchemist labored for centuries to discover the process by which gold was constituted, and just as vainly are some now trying to make diamonds. They are composed of equivalent extracts from the same simple elements from which water is constituted, but by a different combining force, which Infinite Intelligence alone comprehends.

But over the simple elements of sound man has as perfect power of combination and organic control as Infinite Wisdom has over the simple elements of the universe, and has produced about the same number of specific effects by a change in the combination of them that Infinite Wisdom has produced by a change in the combination of the same simple elements of either of the first four departments, considered under the titles of Intellectual Spirit, Organic Force, Celestial Substance, and Terrestrial Matter; the sequence of which is, that the number of simple elements is equal, and that Infinite Wisdom has exhausted the difference in specific effects to be produced by a change in equivalent combinations.

Musicians have also based their science of music on the same axiom of seven simple sounds, and the change of harmony which has already been produced is beyond the comprehension of any one mind, and when they are exhausted will agree in number with that of language, and both together with the other four departments.

The diatonic scale used by all composers of music contains eight notes, representing a rising and falling scale, the first seven of which represent seven simple ascending sounds, and the eighth is a repetition of the first, an octave above.

By a change in the combination of these seven simple representatives of sound all vocal and instrumental music is produced, and the song of every bird may be represented, learned, and performed by an intelligent musician. Thus we discover that the finite intelligence of man has the same synthetic power of producing a change, in effect, by changing the equivalent combinations of proportional quantities from seven simple sounds, that the infinite intelligence of God has in combining all the other simple elements of the universe.

CHAPTER VI.

DEPARTMENT OF LIGHT.

The next department in the great division of universal philosophy, which we have been led into for the purpose of analogical demonstration, and with which the science of sensibility is very intimately connected, is light, which undoubtedly has a specific group of simple elements as distinct from the other five as celestial matter is from terrestrial, and as organic forces are from spiritual; and upon consulting science as to the number of those simple elements, we find the following axiom, (see Comstock's Philosophy Revised, page 276:)

"If a ray of light be admitted into a dark room through a window-shutter, and allowed to pass through a triangular-shaped glass called a prism, the ray will be decomposed, and instead of a spot of white there will be seen on the opposite wall a most brilliant display of colors, including all those seen in the rainbow, which are clearly defined by distinct lines, apparently divided by their density as follows, viz: At the bottom, and apparently the densest, red; next in density, orange; next, yellow; then green, blue, indigo, and violet."

Thus science demonstrates that a ray of light

is composed of seven simple specific colors, in which there is no more light, in their simple uncombined state, than there is in a lump of charcoal, until brought within the synthetic influence and combining force of the sun, whereof, by an equivalent mingling and mixing and combining of equivalent parts from the seven simple colors, a flood of light is produced which radiates into space and enlightens the universe. Those seven simple colors, in their uncombined state, are a mist of total darkness, and as imperceptible to any of the five senses of a terrestrial animal as the molecules of matter or celestial substance, mite of organic force or atom of spirit; but by an artistic combination, which is just equivalent to an illuminating ray, the universe is able to demonstrate itself to the knowledge of souls through the sense of seeing, which is one of the principal mediums through which any soul acquires knowledge. These rays of light, in coursing through the immensity of space, are at length, by some imperceptibly slow process of analysis, finally decomposed, and return to their simple opaque elements, in which condition their relations to each other are repulsive. Consequently they spread out to equalize their department of color elements in universal space; and as the sun is making an eternal effort to create a vacuum of simple colors in its vicinity, they rush off in that direction to counteract it, where the consumption is so great, that a movement in that direction of the simple elements of light is going on from the extremities of universal space, and, like all other great bodies put in motion drawing to a common center, they become compressed at that center, so that there never can be any lack of simple elements for the manufacture of light at the great orb of illumination.

The repulsive nature of all simple elements renders them antagonistic to space, and however few there are of them in a bottle, they will place themselves at an equal distance apart till every nook and corner is filled in equal proportion; therefore it is presumed that there is no vacuum in nature, and that each of the specific elements of the six grand divisions pervades universal space in the most equitable condition, keeping up an uncompromising hostility to vacuum; consequently no useful organization could be accomplished in any one of the departments without an overpowering force, controlled by a sovereign will, for the accomplishment of a fixed purpose, designed by a process of reasoning from cause to effect, by an intellectual spirit operating on the sensational machinery of the mind of a living soul, made up of a celestial body, organic force, and intellectual spirit.

These simple colors pervading the atmosphere become impregnated in the combining masses of all substances and become a fixture, a sort of prisoner in a foreign substance, and when a ray of light penetrates to the place of their confinement, the color reflects its fellow from that ray, and thus is able to demonstrate its presence, and thus the whole art of painting consists in combining paints which hold these simple elements of light in combined proportion to reflect just the colors to produce the effect designed by the artist; and all that exquisite beauty of colors in the floral kingdom is produced by the blending of equivalent parts from these seven simple elements of color in the growth of the plant; and the color of precious stones and earthy substances of every kind owe their peculiar colors to the specific simple element of light infused into them during the process of solidification or crystallization; and the color of blood, flesh, skin, and hair of animals depends entirely on the equivalent proportions of the simple colors secreted by the lungs and communicated to the blood during the act of breathing. Thus these simple colors in combination with material substances are mere reflectors of their own color from a ray of light, but when combined by the sun they become an illuminating beam, and are radiated to make the location of their fellows, which have become substantially combined, visible.

We have now briefly glanced at five of the great divisions of the universal elements of nature, commencing at the base and working our way up to the head of Infinite Intelligence; in doing which we have discovered that substantial bodies are of two distinct natures, being combined from two different series of simple elements: one from terrestrial, the other from celestial elements; and that one is a terrestrial body subject to decomposition by chemical analysis, and that the other is a celestial body on which chemical solvents have no effect.

We have also found that organic forces are combined and constructed for useful purposes from two distinct series of simple elements: one a simple converter of fluids into solids by the rapid process of congealation or crystallization; the other a vital force of growth, building up its structure from a germinating seed to a magnificent tree of a thousand years' growth, and from an impregnated ovoid to a powerful animal, requiring from one to twenty years to complete the structure. So that we are now prepared to investigate the axiom demonstrated by the following chapter, viz: That the simple elements of spiritual intelligence are divided into two series, finite and infinite.

CHAPTER VII.

FINITE AND INFINITE.

As all the elements of nature included in the five departments treated upon in the foregoing chapters are repulsive in their relations to each other, that determines their status as infinitesimally finite, as each elementary atom of spirit, mite of force, and molecule of substance places itself at an equal distance from its fellows, in everlasting antagonism to vacuum, and eternal hostility to any combination that would be likely to produce it, it necessarily follows that no body, ponderable or imponderable, could be formed by spontaneous combination or mutual attraction for any purposes in those departments: consequently the great work of creation must have had its origin in another department, which, in our present mode of treating the subject, falls under the head of department number seven.

The series of elements in this seventh department, in their relations toward each other, are all coherent, and spontaneously combine themselves into sensational lines and a spiritual focus, converging from the extremity of natural elements to a mathematical point, in a similar manner to that represented by the rays of the sun in illuminating nature by radiating lines from the sun's focus.

The cohesive tendency of these elements to spontaneous combinations into sensational lines, extending from a focus coextensive with all natural elements, in unbroken lines of nervous sensation, constitutes their status, infinite, unbroken in lines, unbounded in extent, unvarying in sensational excitement to useful industry and profitable development of the resources of nature, by presenting an infinite knowledge of the quality of each simple element at the focal eyeball, and keeping up a constant impression on the retina, of the infinite resources of nature, waiting for an Infinite Intelligence to develop them.

Without an intelligent head at the focal terminus of this nervous system, to investigate, design, will, and execute, it was of no more use in nature than the embryo pattern of a man while confined in the testicles of its sire, where it is formed.

That infinite system of sensational lines and spiritual focus was the fecundity of a personal representative of Omnipotent Infinitude, and all that was necessary to produce it was an ovoid filled with just the proper substances from the terrestrial and celestial departments of simple elements, out of which the body could be formed.

This infinite fecundity embodied in itself the nature of both sexes for the production of its personal representative, and as the mother gathers from the aliments of nature the specific material for the embryo of her offspring to commence life in, and condenses it in an ovoid in her ovarium, ready for impregnation by the male at the proper period of maturity, so this infinite system of nerves, being in communication with every simple element of nature, gathered the substance and formed the infinite ovoid at its focal ovarium, and at the proper period of maturity the combined force of personal growth and Infinite Spirit of Intelligence of the department having been formed into an infinite embryo, entered into the ovoid and commenced the work of constructing its own form in a similar manner to that in which the finite embryo of a man commences the construction of his body in the mother's ovarium, and worked it up to the stature of a God by the slow process of growth, very much in the same way that the fœtus of a man works his soul up to the estate of manhood.

Man is nourished and cherished through his embryo and infantile state by the paternal care and veinous feeding of his mother; God, by the fecundating nerves of his own system: and having arrived at the stature of a God, he was inevitably an infinite worker, being endowed with an infinite capacity for investigation and understanding of nature's elements, designing of patterns and equivalent combinations to fill them, omnipotent power to execute, and infinite resources to work with, he commenced his work of creation, that will never cease till the resources of nature are exhausted.

It is not necessary that the person of God should be very large. Five feet seven inches in height, and one hundred and forty-four pounds in weight, is about the average size of distinguished men—men who have made themselves distinguished by their own mental powers and intuitive genius; and that is supposed to be the size of the person called Christ, who was the son of Mary, the wife of a carpenter, and worked at that trade till he was twenty-seven years of age, with only such knowledge of human literature as apprentices and sons of poor mechanics were able to acquire in those days, who, by the power of exemplary preaching, and the help of a few poor illiterate fishermen and laboring men, employed in promulgating the same truths, revolutionized the religious dogmas and worship of mankind, prostrated all temples erected to the worship of heathen gods, and built upon their ruins a multitude of temples to the worship of his Father and our God, and destroyed kings, overturned dynasties, and dealt a death-blow to human sovereignty, which is just ready to tumble into the grave of oblivion, and established the epoch from which time is reckoned.

When Philip said unto him "Shew us the Father," he answered, "Have I been so long time with you, and yet hast thou not known me, Philip? he that hath seen me, hath seen the

Father; and how sayest thou *then*, Shew us the Father?" The axiom that like father like son, and that in the son we see the specific form of the father, is too well established to be doubted by any body except some natural, who still believes the world is flat. That he was what he represented himself to be—the verily begotten Son of God—is substantially sustained by thousands of temples in cities, churches in hamlets, and meeting-houses at cross-roads throughout America, Europe, and Africa, dedicated to the worship of his Father, and from which the doctrines which he taught are proclaimed every Sabbath, where, at the time of his birth, they were governed by pagan dynasties, and the only temple of religious worship was dedicated to pagan gods and devouring beastly deities, and daily sacrifices of men, women, and children were made, all of which tumbled down at the feet of his successors; and the entire heads of governments now profess to believe in him, his doctrines, and his Father, as their only God.

For any man to assert that a poor illiterate carpenter and a few illiterate fishermen could have done this, except they were backed up by the omnipotent power of Jehovah, is simply preposterous, and still more so to say that God would have sustained an imposter who came into the world with a lie in his mouth, and declared himself to be the begotten Son of God when he was not.

They who saw Christ saw the specific form of him whose intellectual spirit is the concentrated force of all the intelligence in the sphere of infinitude; and in the form of the person manifested in Christ they saw the personal form of him who sitteth on his throne, which is the focus of the infinite system of sensational nervation through which every simple element in nature, even the "sanguine discs" which course through our veins and thoughts in our minds are constantly before him: and this is the reason why Christ said, "But thou, when thou prayest, enter into thy closet, and when thou hast shut thy door, pray to thy Father which is in secret; and thy Father, who seeth in secret, shall reward thee openly." Everything is present with God, as the tower or mountain and star in the far distance is present in the retina of our eye, and as the firing of the distant cannon is impressed on the drum of our ear.

This infinite system of nervation not only communicates intelligence from the boundary of natural elements to the intelligent head, but they transmit the fiat of Jehovah to the planetary forces, and control the laws of nature, as the nerves of a man control the movements of his muscles; and the former forces every movement of the planets, as the latter do the movements of the limbs of the body. And as the nervous system of a man is connected with and has its focal terminus in his brain, extending from thence in unbroken lines to the very extremity of every part of his body, and his spirit alone can understand their communications or work them to any purpose, so also this infinite system of nervation is connected with and has its focal terminus in God's throne, extending from thence throughout the immensity of space, and pervading the entire simple element of nature, and his Holy Spirit alone can understand their communications, or dictate any movement through them.

They are as essentially his property, and a part and parcel of himself, as the nerves of a man's body are of himself, or the rays of the sun are of that orb; and no person except himself, not even his well-beloved and only-begotten Son, can have any knowledge of the intelligence which they transmit, except he reveal it to him. Hence Christ said, "But of that day and *that* hour knoweth no man, no not the angels which are in heaven, neither the Son, but the Father."

The official station which the Son was appointed and chosen to fill is that of king over the animal kingdom, and his duties are those of a chief justice over all the people of the universe, both in their terrestrial and celestial states of existence.

Judging between man and man by the law of full restitution and a just retribution, which the Father established when man was created, and while the Father controls the elementary spheres and systems of nature, the Son reigns supreme in the dynasty of man, not only in the celestial kingdom, but also among the inhabitants of the earth, which, though not yet established, is only a question of time; and as the worship of the Father has succeeded to the worship of devouring beastly deities and stopped the daily sacrifice among men, and the myriads of churches now dedicated to his worship have been erected on the ruins of heathen superstition and pagan temples of human sacrifice, so his kingdom of the New Jerusalem will be established here on earth upon the ruins of devastating human sovereignty, plundering kingly dynasties, ravaging imperial families, plundering republican rulers, and pilfering democratic majorities; and will put an end to robbery without restitution, bloodshed without retribution, and fraud and trespass unredressed. The whole fabric of political frauds, prerogative lies, imperial and kingly ravagings, will be swept away, and the truth, love, mercy, and justice of Christ's kingdom will succeed it, and every man protected in safety under his own vine and fig-tree: and the accomplishment of that purpose of God is not as far off as most people imagine, and those who accept it and fall upon it shall be broken off from all old political associations and ideas, and accept the new order of things as it is presented to them without comment or amendment, and enter the new kingdom as a little child goes to his primary lesson; but those who oppose it and fall under it shall be ground to powder.

When this is accomplished the Father will control the national order, judicial system, and executive dispensation among men through the administration of the Son and his appointed representatives, as he now does the elements of nature and the spheres and organic bodies which his own fiat has made through the administration of the Holy Spirit which pervades his throne and the infinite system of lineal nervation which centers in it, and thus we have a resonable solution of the Trinity, the necessity of it, and the work assigned to each of the persons, and God is the all in all managing head of it.

As the personal force of growth in the soul of a man has its seat in the heart, and performs its work in a manner independent of the direct and immediate attention of the intellectual spirit, but in perfect harmony with all its ideas, so does the Holy Spirit of the infinite universe of God have its seat in his throne, and performs its work of controlling the planetary system without the immediate attention óf the Intellectual Spirit, but always in infinite harmony with his fiat. The galvanic force is a transmitter of finite intelligence through a finite line of communication. The Holy Spirit is a transmitter of infinite intelligence through an infinite line of nervation. The force of personal growth seated in the heart of a man is a transmitter of finite sensation through his nervous system, at the instigation of his finite spirit of intellect, which has its seat in the brain. The force of the Holy Spirit of universal growth of God's universe, seated in his throne, is a transmitter of infinite sensation through an infinite system of nervation in perfect harmony with his omniscient ideas.

The radiating sunbeams are a finite enlightener of the finite souls of the animal kingdom. The converging lines of infinite sensation open up to God an omniscient knowledge of all the elements and resources of nature. Man is a microcosmic of the earthly planet on which he begins his existence, partaking of the nature of five finite elements. God is an epitome of universal nature, and not only partakes of the nature of all the elementary departments, but is the organizer and controller of all ponderable bodies and imponderable substances, and with his infinite system of nervation permeates all things, even to the very boundary of nature's elements; and having an omniscient knowledge of the effect of all the changes which can be made and the mode of making them and the propensities of all animals, he knows the future omnisciently, while we only know the past and present finitely.

Thus it is evident that there is an immensely

greater difference between that infinite spirit which endows God with omniscience and that intelligent spirit which gives finite intelligence to man than there is between terrestrial matter and celestial substance.

There is nothing any more improbable or mysterious in the axiom of this infinite system of lineal nervation which endows God with omnipresence, omniscience, and omnipotence, than there is in the well-known fact that the rays of the sun endow that orb with a capacity to illuminate the universe to the same extent; and it is just as preposterous to say that all this infinite variety of sun, planets, satellites, stars, comets, mineral substances, vegetable vitality, and animal life could have been created without an intelligent head to dictate the work, as it is to say that the rays of light illuminate the universe without a sun to radiate them, a railroad built without a constructor, and trains run without a conductor.

From the foregoing mode of treating this subject it appears that the simple elements of nature are divided into seven grand departments, and that six of the departments are divided into two serial divisions, which being omnigenously arranged, stand thus:

SCIENCE OF INTELLIGENCE.

Simple elements.	1. Omniscient Spirit. 7	Simple elements 7
7	2 Force 6 of Nervation. | II. S. | Planetary.	7
7	3 Light. 5 Natural. | | Artificial.	7
7	4 Sound. 4 Artificial. | | Natural.	7
7	5 Finite Spirit. 3 Intellectual. | | Sensational.	7
7	6 Finite Forces. 2 Growth. | | Organic.	7
7 7 Cel estial.	Substances.	1 7 Terrestrial.
42		49

Altogether ninety-one simple elements in nature, out of which all organic bodies, whether of spiritual sensibility, vegetable vitality, planetary system, twinkling stars, sound, light, and even God himself, have been constructed; all of which, in their uncombined state, are an ethereal element. Even the diamond is a gas, which to our terrestrial senses is totally invisible, which led the ancients to believe that all visible substances had been created out of nothing; which Christian science has discovered to have been constructed from ninety-one invisible simple elements, classified into seven departments, each department into two branches, and each branch into seven simple elements, as represented in the above omnigenous scale.

In the creation and controlment of the inanimate universe God's executive agent is the Holy Spirit, consisting of two forces: one the force of sensation, transmitting intelligence and orders through the infinite system of lineal nervation; the other the force which controls the planets and leads them through their orbits.

In the control of the political organizations and moral conduct of the animal kingdom, Christ, his only begotten Son, is his chief executive; so that science has vindicated the doctrine of the Trinity, for the Father, Son, and Holy Spirit are a unit in the government of the universe.

Man, too, has three estates of existence to pass through before he arrives at his final estate of manhood, called heavens. The first is his earthly estate, in which he begins his existence; the second is his celestial estate, in which he becomes thoroughly renovated from all terrestrial matter and immoral tendencies; third, the estate of infinitude, where he can see God as he sees his fellows, and enjoy the full fruition of infinite order and human enjoyment. Concerning this matter the Apostle Paul testifies as follows, viz:

"I knew a man in Christ above fourteen years ago, (whether in the body, I cannot tell; or whether out of the body, I cannot tell: God knoweth;) such a one caught up to the third heaven. And I knew such a man, (whether in the body, or out of the body, I cannot tell: God knoweth;) How that he was caught up into paradise, and heard unspeakable words, which it is not lawful for a man to utter."

If this testimony of Paul was false, it is a most

extraordinary case of perjury; for the only reward offered him in this world was bonds, stripes, imprisonment, and crucifixion.

Even for the plain Christian doctrines which he had ventured to preach, he testifies of his reward as follows, viz:

"Are they ministers of Christ? (I speak as a fool) I *am* more; in labors more abundant, in stripes above measure, in prisons more frequent, in deaths oft.

"Of the Jews five times received I forty *stripes* save one. Thrice was I beaten with rods, once was I stoned, thrice I suffered shipwreck," and many other perils which he enumerates.

This same Paul, before he was converted to Christianity, was known as Saul of Tarsus; was educated at the feet of Gamaliel; was a popular orator; held a lucrative office in the government; was venerated by the populace and respected by princes; was very zealous in persecuting Christians, and even held the garments of subordinate officers who stoned them; and when Saul turned Christian, and commenced preaching the doctrines which he held a commission from the government to persecute, he knew that his reward in this world was confiscation of property and loss of reputation—persecution in the precise manner that he had suffered; and certainly nothing but a visit to the third heaven and a high commission in a kingdom which extends beyond the grave could have induced Saul, the persecutor, to become a preacher of persecuted doctrines. If Saul had no security for a reward in the third heaven, it was the most extraordinary change that any educated popular orator with a large income ever made. Such men change for gain, not for stripes and crucifixion. To have spoken what we are now writing would have been certain death to Paul, such was the power of human sovereignty at that time. Verily the kingdom of Christ on earth was compelled to develop itself like a grain of mustard seed—first a shoot, then a leaf, then branch after branch, till the seed is now ripe and ready for the harvest. Paul had a commission from Christ himself, and there are many others to be bestowed not many days hence.

A net-work of railroads for the equalization of commerce and providing a uniform supply for every demand of the animal kingdom over the face of the whole earth is being spread out over it with the same wonderful and mysterious progress that a net-work of veins and arteries is spread out through the body of a man for the purpose of supplying every portion of it with just the substance needed for its construction and repair, performing the same function for the corporeal person of a man that the net-work of railroads is destined shortly to perform for the corporeal kingdom of Christ; and the establishment of the heart and focal terminus of it is not far off.

A system of telegraphic lines has nearly encircled the world in the same mysterious manner that the nervous system is made to pervade a man's body, which will soon find their terminus in the head of the kingdom of Christ, leading straight out from thence, and pervading every precinct and school district among men over the face of the whole earth, so that each individual in the kingdom can have as instant a communication with the head of the government as the extreme end of a toe or the finger of a man has with the head of its own body; and an injury done to any person, child, or animal will be as certainly known at the head, and restitution and retribution made and enforced, as any injury done to any part of the corporeal person of a man is known at the head, and reconstruction performed by the force of growth from the seat of life.

The spiritual head of this corporate personification of divine government among men is Christ the only begotten son of God, and the rule of law by which justice is to be dispensed among all people is that law which God inscribed upon the tree of knowledge of good and evil in the Garden of Eden, and afterwards upon the tablets of stone on Mount Sinai, which Christ confirmed in his first sermon and lived in perfect fulfillment of, and which Judge Blackstone calls the law of revelation, and is universally known as the law of full restitution for all property damaged and a just retribution for all personal injuries, based upon loving thy neighbor as thyself and the Lord thy God with all thy heart.

The whole machinery and function of this new order of government is described in the new book of the law that is shortly to make its appearance in the world, which when John saw it was sealed with seven seals, the last of which is being broken, and once opened will explain the mode of administering the government so clearly that a child could not err in dispensing justice by it, before which human sovereignty, with all its machinery for robbery and plunder, will fall down and be broken to pieces as Dagon fell down and was broken before the ark of the covenant containing the tables of stone on which the primary maxims of the divine code of law were inscribed, and this heavenly machinery of government, once set in motion, will crush out and destroy every other form of government which human sovereignty has invented, as the holy doctrines of Christ did pagan idolatry and human sacrifices to beastly deities, and the heavens of human sovereignty, together with all their lying statutes by which courts work out robbery, and the earth of aristocratic prerogatives with all their diabolical privileges, will be rolled up like a scroll, burned with fire, and the ashes scattered by the four winds of heaven, justice, mercy, equity, and truth, and the kingdom of Christ established in their stead as John the Relater saw it, when Christ will be reigning King among men, and God all in all.

Under this régime the same peace, harmony, and good order will prevail throughout the earth among all people that now prevail in the planetary system, and each and every man will stand upon a common equality before the maxims, statutes, and judgments contained in the new book of the law, and every man will dwell safely on his own homestead, reap according to what he sows, and enjoy the full benefit of his own labor.

CHAPTER VIII.

REMARKS ON INFINITUDE.

In the foregoing brief consideration of the seven departments into which the simple elements are naturally divided, the attentive reader has discovered that the inherent relations between the simple elements in the sixth and seventh departments are antipodal to the inherent tendencies of the simple elements in the first, second, third, fourth, and fifth departments.

For in the department of substance, force, spirit, sound, and light, the specific tendencies of the simple elements in each department are repellant: *i. e.*, each molecule of substance finds its natural position at a mathematical point, which places an equal distance between every molecule in the universe, in uncompromising hostility to vacuum; so that throughout the boundless expanse of nature each molecule is at an equal distance from its next neighbor. As gas confined in a bottle—however little there may be of it—pervades and fills the whole interior, so do the molecules of matter and substance pervade and fill the boundless expanse of nature, so stationed that the distance between each molecule and the mathematical point at which its next fellow is located is exactly the same in every direction, which leaves plenty of room for each mite of force, atom of spirit, particle of sound, and speck of light to distribute itself in the same mathematical order of uncompromising hostility to vacuum; and all thus arranged still leaves an uninterrupted passage to the rays of light, vibration of sound; and the omniferous mass being permeated by the omniscient system of sensational lines, constitutes God an omnipresent, omniscient, and omnipotent personification of Infinitude.

Thus we discover that the simple elements in the five lower departments of nature, viz: substance, force, spirit, sound, and light, on account of their inherent repulsiveness, fill the infinite expanse of nature in detached particles; while the simple elements of the seventh department have an inherent affinity for each other which constitutes them into a nervous system of infinite dimensions, all terminating at the person of God, which is so absolutely fixed and established by their own inherent affinity, that even he cannot change the fixed terminus, sever a sensational nerve, or make the least alteration in its organic structure. That whole nervous system is as much a part and parcel of himself as the nervous system of a man is part and parcel of his own soul; and as his finite system is under the control of his mind and subject to his will in its complete organization for useful purposes and the proper enjoyment of life, while he has no power to change its organic form to the breadth of a hair in any particular, so also it is with that infinite system of nervation, being a part and parcel of the person of God; as mysteriously constructed by the Holy Spirit of personal growth as the nervous system of a man is by the finite spirit of animal growth ere his reasoning capacity was developed. He has no more power to change that infinite system of nervation than man has to change his finite nervous system, while through its infinite permeation of all the elements of nature he obtains omniscient knowledge of all its resources, and omnipotent power to develop them and produce every effect which any change in the combination of the simple elements can produce. And this infinite system is as subject to the movements of the person of God as the finite system of a man is to the dictation of his head. Wherever God turns the infinite nervous system turns, and wherever he goes he is still the focus of it: even the entire machinery of his throne seems to go with him.

Ezekiel, in one of his prophetic visions, saw it once at the River Chebar, and described it in the 1st chapter of his book, and again at the temple in Jerusalem, and described it in the 10th chapter. Wherever God chooses to go through the immensity of his works, he is still the focus and personification of the omnipresent system of nervation—the sensational heart of constructing genius, the intelligent head, in which the device of every substantial form is modeled, and the pattern of every creature and thing which ever has or ever will demonstrate itself is drawn. By a careful organization of the simple elements, nature is drawn and stereotyped into fixed stars, migrating comets, revolving planets, or illuminating sun, sparkling gems, specifically formed crystals, massive rocks, barren sands, alluvial soils, and vegetable-producing loams, each specific vegetable bearing seed after its kind, and every species of living creature to repeat itself by copulation; and wherever he may be, in his person is concentrated the infinite system of creative nervation which fills the entire department of the seventh sphere of infinitude. For God to change his personal location from point to point throughout the immensity of space has no more effect on the perfect working of his infinite system of nervation than a change of location by the soul of a man does in the working of his finite nervous system. The soul of a man is always the center of action for his nervous system wherever it may be: so the person of God is also the center of action for the infinite system of nervation which fills the entire department of infinitude wherever he may be; and the Holy Spirit of the personal growth of his person has its seat in his heart, always working in perfect harmony with the spirit of omniscient intelligence, which has its seat in his head: whereas the planetary, mechanical, driving forces of the Holy Spirit, which fill the entire department of the sixth sphere, have a fixed central focus, from which the whole machinery of the planetary system is worked, and the unvarying movements thereof through their immutable orbits kept up, which are as stationary as that of the sun, from which the universe is illuminated.

This focus of the sixth department may very properly be termed the heart of universal planetary movement, where the organic and mechanical driving force of the Holy Spirit has its seat, and executes God's fiat of planetary construction, station of them, and movement in their destined orbits, and bears a similar relation to the working machinery of the planetary system and various movements of the heavenly bodies that a steam engine does to the working of the pulleys, looms, shuttles, and spinning-jennies in an extensive cloth manufactory. It is the machine-shop and engine-room of planetary movement, and is the laboratory of Jehovah—God's home and usual place of residence, where his business-office is located, and is the capital of the universe, which inspired writers have termed the City of the Living God.

Thus we discover that the sixth department is simply the laboratory of Jehovah, in which the cohesive, pliant, constructive, and driving forces of the Holy Spirit work out his design, having its seat of operations at the engine-room of the universal manufactory in the centre of the City of the Living God, in the midst of the sixth heaven; while the seventh is completely filled by the infinite system of nervation, which in the beginning was the fœtus which produced its own personification of infinitude, and constitutes him the omnipresent, omniscient, and omnipotent organic controlling head of all the elements of nature, in whom the concentrated spirit of omniscience works out the universal harmony of Infinite Intelligence, who in the beginning was the only person who had joy in mental device, or could rejoice in the work of his own hands.

Not wishing to monopolize all the enjoyment which intelligence alone could produce, and having omniscient knowledge of all the variety of enjoyments which could fill the universe, by peopling the planets with animate creatures, who were conscious of their existence and had sufficient intelligence to know their own thoughts, in due time commenced the development of the simple elements of the finite spirit of intelligence and finite growth, which began with a monad and terminated with the construction of man, who, as they multiplied upon the face of the earth, spent much of their energy in devouring each other: to counteract which and organize them into one fraternal family, in which each one understood that by working for his own best interests he could do nothing to injure his neighbor, and therefore advanced the general good of the whole. In pursuance of his own purpose, ere forming man in the beginning, he prepared a reigning head of the whole family in the person of Christ, who, being legate of God, was filled with the Holy Spirit of his Father's intelligence, and born of a woman, partook of the nature of man, as all of the substantial elements of his soul were drawn from the departments of finite substance, in which he feels all the infirmity of the flesh and concupiscent desires; but having the Holy Spirit of the Father to direct and control his actions, he set an example of a holy life for the emulation of the family. For this purpose, God chose for the mother of his only begotten Son a virgin, from one of the poorest families in Judea, in which he worked his way up to manhood at the occupation of a carpenter, and suffered all the indignity which the rich families, professional men, and officials could heap upon him, and all the pain which human sovereignty could inflict by a death upon the cross, so that he did suffer all that man can suffer by the tyranny and oppression of his fellows, who having usurped that sovereignty which belongs only to God, is certain to become a satanic tyrant, and from this sad experience in his estate of manhood, knows exactly what organization of government is required to produce millennial order in this world, and establish peace among all the nations and families of the earth and good will among men, and the statutes, judgments, and mode of redressing wrongs, which will maintain it. All of which is written in the new book of the law, which when John saw it was sealed with seven seals, which Christ alone had power to unloose, the last of which is about being broken, and his kingdom established on earth as John saw it; and the status of the third person of the Godhead accomplished, the sublime machinery of which will enforce a frugal administration of it, and place every man upon a common equality before the law, of full restitution for all property offenses and a just retribution for all bodily injuries, which are among the fixed purposes of God the Father to be accomplished through the agency of his Son Christ Jesus our king, that no man can stay for a single moment beyond the appointed time; and he who does not see that its accomplishment is near at hand does not observe the signs of the times.

Thus from the inherent affinity of the seven series of simple elements in the seventh department of nature to constitute themselves into an infinite fœtus for the production of an omnipotent person of omniscient intelligence, working itself up to the estate of an omnipotent God, in a similar manner to that which the finite fœtus of man works itself up to the estate of manhood, so that the whole elements of the department of infinitude being as effectually constituted into an omnipotent person as the fœtus of a man develops itself into one finite person, therefore that is the end of self-organization: for the reason that outside of that department the inherent status of all the simple substances of nature are repellant, so that an overpowering organic force is the only means of producing any combination among them for useful purposes.

Having thus looked through nature up to nature's God, and discovered an omnipotent person with an omniscient head to devise and an omnipresent heart to execute, with infinite elementary resources to develop, we are now prepared to enter the field of finite intelligence, and investigate more particularly the simple elements necessary to constitute a finite soul, and learn what human reason will enable us to discover concerning the manner of their combination.

CHAPTER IX.

SIMPLE ELEMENTS OF SOUL, SPIRIT OF INTELLIGENCE.

The seven simple elements of finite spirit, which in the foregoing omnigenous scale stands department No. 3, ascending from substance to infinitude, and No. 5 in the descending scale from infinitude to substance, which is the department of simple elements, from which are drawn all of that spirit of intelligence which gives mobility to the animal kingdom, and endows each species with a specific degree of intelligence, may very properly be designated as follows, viz:

Spiritual forces.			DIATONIC SCALE.		Mental faculties.	
"	"	Ideality.	B. No. 7.	Invention	"	"
"	"	Memory.	A. No. 6.	Recording	"	"
"	"	Will.	G. No. 5.	Executive		"
"	Judgment.		F. No. 4.		Decision	"
"	Discernment.		E. No. 3.		Choice	"
Consciousness.			D. No. 2.			Sensation.
Impulse.			C. No. 1.			Mobility.

The above specified seven simple elements which fill the 3d department of the seven grand divisions into which the simple elements of the universe are naturally divided may very properly be termed the alphabet from which all the poetry and prose, joy and sorrow, felicity and anguish, enjoyed and suffered in the omniferous chain of animal life is composed, and perhaps more properly the diatonic scale of characteristics from which the harmony of all the intelligence which is manifested thoughout the animal kingdom is composed. For the status of each species of living creature, which is endowed with any degree of mobility, depends entirely on a combination of proportionate equivalent numbers, selected from the above designated simple forces. For each living creature on the face of the earth, from man to a monad, does demonstrate by his specific habits and individual acts that he has more or less of the element from each of the above series in the composition of the spirit of life which endows him with mobility.

1. IMPULSE.—The function of impulse, in its combination with other spiritual elements which help to constitute the spiritual pattern of a species of animal, is mainly that of the fingers of the spiritual operator, with which the delicate touches of sensation are given to the keys that operate the nervous system, puts the muscles in motion, gives force and direction to every movement of the body

and utterance of every word, and is under the entire control of the will.

As the fingers of the telegrapher work the operating lever on a telegraph machine, by which the message is sent on its mission, so also the spiritual finger of impulse works the dictating nervous leverage which sets the muscles in motion for a bodily movement. And as the fingers of a skillful musician give the exact touch to the keys of a piano to produce the proper sound in a set piece of music, so does the spirit's fingers of impulse touch the keys of the nervous system to produce the proper sensation in chord with the set harmony of the soul.

Impulse is an essential element in the organization of the spirit of every species of animal. As no movement of a muscle could be produced without it, consequently every living creature that moveth by volition has at least a small extract from this simple element of nature mingled in the combination of its spirit; for without it the nervous system would be of as little use as a piano without the fingers of the musician.

2. CONSCIOUSNESS.—The simple elements of consciousness in the spirit of an animal give it the knowledge of what passes in its own mind, consequently of its own existence and of surrounding objects, but give no idea of any difference in objects or aims in life; still, without some small portion of these elements in the combination of its

spirit no animal could exist, for it would not know when it was hungry or that food was necessary to sustain it. Consciousness is the lowest grade of the series of spiritual elements, a connecting link between animal life and vegetable vitality, and constitutes the first grade in the ascending scale from vegetation. Still it is an intellectual element, an essential ingredient in the spirit of sensibility: its function is in the brain, and would be of no more use, combined in the force of animal growth, which has its seat in the heart, than an eyeball would be if placed in the bottom of the foot. The floating specks of consciousness are never of any use, except when combined in a spirit of some species of animal which has a nervous system capable of being set in motion by spiritual influences.

3. DISCERNMENT.—Discernment is the simple element of finite spirit, which enables the spirit in which it is mingled to understand the nature of surrounding objects; to draw a nice distinction between their qualities; to separate and classify them according to their proper relations, and readily comprehend the nature and fitness of things.

Without some portion of this element in its spirit, no animal could distinguish between poison and wholesome food, friend and foe, good and evil, and its life would necessarily be very short upon the earth; for if by chance for a brief season it escaped death by poison, some carniverous adversary would be certain to gobble it up. This element is the main spiritual ingredient in the soul, which gives it a proper degree of caution for self-protection; and weak timorous animals, such as deer, rabbits, and inoffensive birds, have a predominating quantity of these last three elements of spirit—impulse, consciousness, discernment.

4. JUDGMENT.—The simple elements in the department of finite spirit designated judgment are the spiritual ingredients of the soul, which give it decision of character, stamina of nerve, and perseverance in purpose. Discernment and judgment are indispensable qualities in a judge, and dogs, horses, and many other domestic animals frequently manifest a greater degree of it in their composition than their owners, and elephants and other wild animals frequently make quite as good use of it as men for the advancement of their specific interest and enjoyment.

5. WILL.—The force of will is constituted by mingling a proportionate quantity of the simple elements from the fifth grade of the ascending series in said third department, as designated in the above table, with the specific elements of a soul in constituting the type of a race, and no animal could execute any purpose whatever without it.

Will is the mainspring of all movement of animate bodies by volition. In all well-regulated minds it is subject to judgment, and acts in accordance with its dictation, but is not necessarily subject thereto, and frequently acts with very little regard to the dictates of sound judgment, and in its unreasonable use of impulse frequently does great injury, not only to its own person, but others. The species of bull-dog, for instance, evidently has a very large supply of spirit from this grade in the composition of the spiritual force which constitutes him a living soul, and a good many men are in the same category, who in their unnatural purpose to destroy others bring destruction upon themselves. Will is the executive force of the soul, and without it no purpose could ever be accomplished.

6. MEMORY.—The sixth grade of simple elements in the department of finite spirit is what gives to any species of animal the faculty of memory, whereby the effect of each action is recorded on the tablets of the mind for a guide in future actions, and is as necessary for the use of judgment as a law library is to a judicial tribunal; and every animal living has more or less of this element in its soul, and for the few records necessary to their welfare, in some of the lower species, their memory is much more retentive than that of man.

Men have such an immense number of events to record, that the tablets of their memory become so crowded that judgment frequently has a hard task to find the necessary record in time to dictate a proper action, even when the will is subordinate.

No person of the slightest observation can deny that every species of animal has memory.

7. IDEALITY.—The seventh and highest grade in the scale of simple elements of spirit is the ingredient in the spirit of any animal which gives it the faculty of invention, and endows it with a capacity for constructing whatever is necessary for its comfort and convenience. Without some portion of this element in his spirit no bird could build its nest, no animal its burrow or habitation, no spider its web; neither could a fish prepare the copulating clean gravel bed for its spawn.

Ideality is not the only originator of ideas; indeed, the largest portion of the ideas of any species of animal springs from a very different source. Consciousness produces an idea of existence, discernment of the nature of things, judgment of their fit use; will appropriates and memory records an idea of the effect of a cause by the use of the five senses, investigating surrounding causes and effects as they exist. But ideality investigates and reports the facts involved in cases of new combinations, and stimulates the soul to make the experiment and test the effect.

The predominance of this element in the soul of man is what gives him his great pre-eminence over every other species of animal; even the inventive ideas of the wild Indian races are as stereotyped to the construction of a wigwam, sweat-house, willow basket, bow and arrow, stone hatchet, canoe, bone spear and elder whistle, skin moccasin and wrapper, as the ape is to its bush tent, the beaver to its dam, or the bird to its nest. But the element of ideality predominates to such an extent in the spirit of man, that the exploring of new fields of combinations of everything which he has any power over, for the purpose of producing labor-saving machinery, various commodities of general utility, sundry ornaments and amusing games, prose, poetry, and song, ethics, religion, and philosophic speculation, so enlarges the circle of his inventive genius, that it will take him eternally to get around it, without any possible chance of exhausting the resources of the finite field in which he is operating.

The superabundance of this ideal element which the Creator mingled in the composition of the spirit which constituted the patriarchal pair of the species of man a living soul, is what drew that marked line of distinction between the nations of men and the Indian tribes, elevating the former to the image of God, in industry and inventive genius as well as in personal resemblance, and places man on a pedestal entirely above the reach of the Indian's inventive vision, and leaves the Indian specific soul on the common level of feeders at the table of nature, plodding vagabonds, merely the head-link in the chain of groveling animal life, with no aspirations above a wigwam for a habitation, and the only furniture therefor a willow basket, which are repeated from generation to generation with the same specific architectural resemblance that the beaver builds its dam, the wasp its nest, and the sea sets up its cone of corals; while man is not only an epitome of the omniferous mass of inventive wisdom, but his inventive genius soars high up into the second heaven of his Creator as a useful worker.

Thus we discover that each of the above series into which the simple elements of finite spirit are divided is a key-note to the specific form, action, and harmony of intelligence of each race, which constantly repeats itself in the specific forms and habits of the patriarchal pair, and that the natural plan or key-note for animalculæ is in C, Indians in B, while the key-note to the harmony of man's intelligence is on B sharp, several octaves above the key of the Indian race in B natural; for every careful observer of Indian character readily discovers that there is a much wider difference between the artistic harmony of the intuitive intelligence and inventive genius of man and that of the Indian races than there is between Mozart's best effort at operatic song and the simple strain of raising and lowering the eight notes: still all of Mozart's pieces are composed by an artistic change in the combination of those seven elements which in their full proper sounds are so simple and uninteresting. So also the superior harmony of man's intelligence was composed by an artistic combination of the same simple elements of finite spirit which constitute an oyster a living soul, and an Indian a come day go day, God send plenty, happy vagabond, enjoying himself in his own simple mode of life, with more contentment than man does in rushing upward and onward in his great work of civilization.

Over the diatonic scale of eight notes, representing seven simple sounds and one octave repeat, man has the power to combine and recombine those simple sounds on major and minor scales, sharp and flat keys, by tones, semi-tones, and octaves, and has already produced myriads of set pieces, with as high as a hundred different parts to be sung in the same piece, and the effect produced thereby in the animal kingdom is as multifarious and varied as the species; but altogether they have but just commenced the work of demonstrating the number of effects which can be produced in song ere the changes in those seven simple sounds are exhausted.

Man of course has excelled in this department, as he does in every other station. Still he is far from being the only musician. Several varieties of birds have mastered a number of octaves, and some of them put man on his best composition to equal the harmony of their strains. Indeed, there are some feathered songsters which have composed pieces and sung strains therefrom which it is doubtful whether man will ever be able to equal or harmoniously imitate with voice or instrument.

Then there is the katydid: what variation, harmony, and volume of sound it produces for such a little body. It is astonishing how such a little head can compose such a variety of harmony, and sing so enchantingly to while away the long dreary nights. Then there is katy's wood-nymph rival, the locust, which has far surpassed anything which the Indian has ever accomplished in the line of musical composition. And who has not been charmed in childhood by the artistic concerts of the crickets under the domestic hearth-stone? And how often in riper years do our minds withdraw from the turmoil of business and go back in imagination to those juvenile, joyous scenes of motherly care and fraternal surroundings, and draw new inspirations of confidence in the protecting care of our Heavenly Father, confidence in ourselves, strong hope of success in the future, from our recollections of the persistent, thrilling, harmonious concerts nightly performed by the crickets under the hearth? Then there is our old favorite the bobolink. In the daisy meadow how often have we been charmed by the sweet warbling of his artistically-arranged sonnets, as he gracefully flitted from daisy to daisy, beating out the time of the tune with his illustrated wings, while hopefully waiting for the appearance of his little family, which his faithful consort was nursing into existence with the assiduity that our own mothers watched over us in unconscious infancy, and sung sweet lullabies to us in the cradle; and thus we can trace an immense variety of effects in the harmony of music which have already been produced in changing the combination of the diatonic scale of seven representatives of simple sounds by the composers among insects, birds, beasts, and men, from the chirp of the cricket up to the best effort of human composers in sacred harmony, operas, and doggerel song; and when convinced that the work has just commenced, we are no longer surprised at the infinite variety of specific effects which God has produced in the harmony of life's intelligent pursuits, as demonstrated by the different moods in which the various races most enjoy themselves; for each species, when left to itself, free from the influence of man, adopts the habit of life and plays out all the harmony of intelligence which the combined spirit of the species is master of in the mood in which its whole soul most enjoys itself; and man does exactly the same thing, and nothing more.

All that man ever has done or ever will do is as completely the effect of the original artistic composition of a set piece of intellectual harmony by a skillful combination of all the harmonious elements in the above diatonic scale of simple elements of finite spirit, designated in the book of animate harmony under the title of Man, as any set piece in the Harmonia Sacra of church anthems is the result of a skillful combination of the seven representatives of simple sounds in the diatonic scale for musical composition.

Thus, while the measure of composition is similar, the execution is antipodal, for the reason that the diatonic scale of eight notes represents inert sounds, and the piece composed thereby has to be performed by living artistic souls; whereas the diatonic series of simple elements of finite spirit represents infinitesimal specks of live dancing spirits which, when combined into a specific spirit for some useful purpose, plays its own part in the choir of animal life, and demonstrates the quality of its own inherent harmony by its intuitive action and voluntary mode of enjoyment.

The scale of eight notes is as readily combined into doggerel songs, jigs, waltzes, and operas as it is into anthems. So, also, the scale of seven simple elements of finite spirit enters as readily into the composition of the specific spirits of fish, reptiles, insects, fowls of the air, and beasts of the field as they do into the composition of the finite spirit of man, who next to God is the highest order of intelligence, notwithstanding many in the past centuries, under the diabolical reign of human sovereignty, have been brought down and degraded to a condition of barbarism far beneath any reptile or devouring beast, in which condition they became veritable devils; and the sole mission of Christ was to remedy this evil, by the establishment of a divine organization of government, so that children will be born in heaven, work out their earthly destiny under a holy organization of government, and pass away to the eternal heavens of God without any change in the machinery of government, rule of law, or mode of dispensing justice, and every animal on earth will be protected from oppression, and dwell safely in their own mode of life.

Each new effect produced in music by a composer is the result of a series of experiments in the arrangement of the eight notes to chord with the preconceived idea of the composer, on account of the limited sphere of his understanding concerning the effect a certain arrangement will produce. Whereas God's omniscient knowledge of the precise effect that every change in the equivalent combinations of proportionate numbers of spectral spirits selected from the seven serial representatives of the above diatonic scale will produce, does away with all necessity for experimenting on his part: so that when he commenced the composition of the specific pieces which fill the book of intellectual harmony, there was no hesitancy in selection or doubt as to the effect which a certain combination would produce. Consequently he commenced with the lower or monadic piece, and set the precise number of spectral equivalent elements from the said spiritual scale in each specific race of monads, to complete the piece in all its parts, with the same unhesitating precision that a compositor sets the representative types in form to repeat the manuscript copy lying before him, till the spirit of the patriarchal force of each species of monad was composed, and the monadic piece set in complete order in the book of intellectual harmony.

Next the spirits of the multifarious species of fish were composed and set in harmonious order; next worms, snakes, larvæ, and insects, and so on through all the specific races of fowls, birds, beasts, apes, negroes and Indians, till the changes are finally exhausted in the composition of the finite spirit of man.

As the spectral elements in the above spiritual scale are all repellant by nature, it necessarily follows that the composition of the spirit of the patriarchal pair of any species of animal could only be accomplished by the omnipotent power of the combining force of the Holy Spirit, which being immutable in all of its works, the status of the spirit thus composed, as to form of body, modes of enjoyment, and specific pattern of propagation, was as unalterably fixed as the earth is in its orbit or the moon in its lunar circuit, and has no more power to change its organic structure than they, or specific force by copulation than they have by their movement, nor to travel out of its fixed circle of intelligence than they have to leave their orbit. Neither has the spiritual type of a species any more power to change the mood

SIMPLE ELEMENTS OF SOUL. 127

of its enjoyments through the endless increase of numbers by copulation to the end of time, if there ever is an end to time for copulation, than an anthem has to change a tone in its composition by being repeated an indefinite number of times by the same stereotyped plate.

The specific mode of enjoyment, stereotyped form of copulation, and individual personal appearance of the whole race of a species are immutably fixed and determined to all eternity by the equivalent proportions of spiritual elements selected from the above scale of finite spirit, and mingled in the composition of the spirit of the patriarchal force of the species, which is sustained and proved beyond a doubt by the constancy with which the whole race of each species has been cast by the copulating mold in the specific form of the pattern of its prototype since the memory of man.

The physical structure, mental moods, and copulating pattern of every race of animals is fixed and determined by the composition of the spiritual harmony set in the spirit of the protoplast of the species, as above described, and the force of animal growth conforms strictly to it in all of its works.

CHAPTER X.

SIMPLE ELEMENTS OF THE FORCE OF ANIMAL GROWTH.

In the foregoing omnigenous table, giving the seven grand divisions into which all the elements of nature are naturally divided, the department No. 2 in the ascending scale embraces the simple elements of nature, designated the finite forces, which are divided into two branches, denominated the force of growth and organic forces. The forces of growth are also divided into two parts, one of which is employed in the production of vegetable formations, and the other in the development of the animal kingdom. The simple element of the forces of nature, as used for the latter purpose, are divided into seven series of constructing agencies, building on four comprehensive patterns of physical structure, animated into an infinite variety of self-propelling bodies, by means of an internal system of leverage, muscles, and nerves, based on seven general plans of physiological construction, each of which is diversified by a great variety of species, quickened into sensational, reasoning, copulating bodies by the variously composed spirits of life, as described in the foregoing chapter. The following table designates the seven general patterns under which the simple forces of animal growth are employed, viz:

Table of seven structural patterns of the animal kingdom.

Soprano.	7. Bipeds,	Mimmifers,	Vertebrates.
Tenor.	6. Quadrupeds,	Mammals,	Vertebrates.
Contralto	5. Fowls,	Fllogan Oviparous,	Vertebrates.
Alto	4. Worms, Larvæ, and Insects,	Articulates,	Cylindrics.
Baritone.	3. Fishes and snakes,	No legs, Oviparous,	Vertebrates.
Bass.	2. Shell-fish,	Radiates, or	Mollusks.
Sub Bass.	1. Animalculœ,		

The above divisions of physical structure of the animal kingdom into seven grades of patterns is very useful in classifying and determining the status of the immense variety of species, as through all the varieties of races in each group there is something of a similarity of physical form, physiological structure of the nervous system, and phrenological arrangement of the machinery of the mind, grade of intelligence and mood of enjoyment, social habits and moral condition, whereby the grade of the elementary atoms of finite spirit used in the composition of the specific spirit is determined; for these forces of animal growth are always constituted in perfect harmony with the quickening spirit of intelligence with which they are associated, and invariably organize the body on the physical, physiological, and phrenological plan which is exactly adapted to the specific pattern, power of action, and mental capacity of the intellectual spirit with which it is associated; sc that, however uncouth it may appear to us, still it is the acme of perfection to the spirit which dwells therein, because it was the pattern of the structure necessarily so formed by the nature of the simple elements of spirit mingled and combined in the composition of it.

The elementary mites, selected, mingled, and combined for the production of the animated specifically-formed bodies in this first grade, are principally confined to transparent substances, which are of so frail a nature that the terrestrial portion of the body requires the constant protection of the fluid in which it began its existence. On being removed from the water, all the terrestrial matter in the body immediately becomes dry and useless, and the soul takes its leave and passes away to its celestial state of existence. Small as these souls are, still they are composed of simple celestial elements, which chemical solvents have no power to decompose or disarrange in the slightest degree. Like every other species of living creature, they are produced by copulation, whereby the simple elements of their souls are forcibly combined into an organic structure, in accordance with the immutable purposes of God and omnipotent rule of reproduction of infinite numbers of the prototype, as established by the constructing force of the Holy Spirit; so that from the birth of each animalcule, its individuality, degree of intelligence, and specific form are as everlastingly fixed and determined as the decree of Jehovah can make it, and its little cup of intelligent enjoyment is as completely filled in its little circle of animalcule ideas as man's is in his great circle of human invention and mechanical progress; and it is perfectly happy in the consciousness of its existence, knowing its own thoughts, and the excitement of seeking out the supplies for its own wants.

As this animalculæ department is the very smallest in size and frailest in physical development in the animal kingdom, we therefore understand that it is produced by the weakest combination of elementary mites from the department of forces of animal growth, and in accordance with the spiritual patterns composed by the lowest combinations of atoms of spirit, and consequently the intellectual harmony is keyed on C natural, and runs all the way through the staff on the lowest lines; still, as it is all they have any knowledge of, they are just as happy as it is possible for any intelligent soul to be, and fill the first measure of intellectual harmony. Had they been left out, the piece would have been incomplete, one sphere of enjoyment in the omniferous heavens unoccupied, God's works unfinished, and one resource of nature undeveloped, whereby a vast amount of intelligent enjoyment would have been excluded from the animal kingdom.

Imperceptible as these little bodies are to our natural vision, still when placed under a microscope we discover several different species, as full of life and animation as any in the higher grades of larger magnitude. They are very active in the pursuit of their food, and the larger species demonstrate their carniverous propensities by gobbling up the small species as readily as a trout does a minnow, and they evidently play as many tricks with each other as kittens or monkeys; and who is prepared to say that the wisdom and goodness of God are not as discreetly demonstrated in filling this animalcule sphere of heavenly enjoyment as in composing the larger links of the animate chain of intelligent life? The physical structure of these animalculæ demonstrates the quality of the mites of force used and the mode of combining them; for each change in combining the simple forces of growth, in composing the soul of the patriarchal type, makes some specific change in the physical appearance of the race, which is just as marked and distinguishable now as it was on the day it was created, and will continue throughout eternity just as specifically the same, through all their numberless increase by copulation, as the same number of quartz crystals would be, or cubes of salt, formed by the natural process of crystallization. And as the intellectual spirit, force of animal growth, and celestial substance are immutably combined in the composition of the soul of each animalcule by the omnipotent force of the Holy Spirit, therefore it can neither change the mood of its enjoyment, physical structure, physiological system, phrenological development, nor a single constituent element of the soul; neither can it cease to exist: once an animalcule always an animalcule, and to its little measure of enjoyment there is no end: for the purposes of God are without recall, and the acts of the Holy Spirit are irrevocable.

CHAPTER XI.

SECOND GRADE.

COMBINATION OF ELEMENTARY FORCE OF GROWTH EMPLOYED IN THE PRODUCTION OF SHELL-FISH.

This second grade in the scale of animal formations exhibits a stronger combination of the mites of force than the grade of animalculæ, but less activity; indeed most of them are without any means of locomotion, and are compelled to resort to strategy for the procurement of food.

There is no pattern in the spirit of these creatures for legs, arms, or even bones of any kind, nor of fins or other means of locomotion, and many of them attach themselves permanently to rocks, logs, and even to the bottoms of vessels, in which condition they make long voyages through strange waters as unconsciously as we make the daily trip of about eighteen thousand miles in the diurnal movement of the earth.

Having no means of escape by locomotion from the multitudes of carniverous fish which surround them, coveting their delicious pulpy bodies, even the present infinite number would be gobbled up in a very short period of time and all their races extinguished and the further increase of their souls cut off, were it not for that strong indigestible shell in which they dwell as securely as a soft-shelled crab does in the crevice of a rock.

Many of these species are covered with a most ingeniously constructed box, composed of two equal halves, with a strong perfectly-fitting hinge, which they can open and shut at their pleasure by a strong muscular movement, which close with such perfect fitting of rim to rim, that even the gastric juices of the stomach of a whale would be completely excluded: so that if they were swallowed with the shell unbroken even by that monster they would pass through the ordeal of digestion and off in the draught of excrement with as little harm as a pearl would stand the same ordeal.

Thus the wisdom and goodness of God are strikingly manifested by the knowledge of secure means of self-protection during the initial stage of their copulating sphere of existence, with which the soul of every species of animal on the face of the earth is endowed by the proper selection of the simple elements of spirit in composing the intelligent harmony of the type of the species to adapt itself to the fitness of the surrounding circumstances in the sphere of its earthly existence, till its destined increase of the number of its species by the process of copulation is accomplished.

Notwithstanding the souls of these myriads of species of living creatures have no knowledge of providing for themselves any means of locomotion, it is undoubtedly made up by their engineering capacity for providing their delicate bodies with a fortification exactly adapted to their station, and of sufficient strength to protect them from all ordinary assaults; and the lack of means to pursue and capture the lesser fish, on which they principally subsist, is abundantly made up by a strategic capacity to draw their food to them; for whenever the sense of hunger is felt, all they have to do is to spring open the supple hinges of the gate of their castle and expose the tempting pulp of their bodies to the teeming myriads of animalculæ that are always swarming about them, which immediately fasten upon it like flies upon an exposed surface of fresh meat, and are immediately secured by the closing of the shell with a snap somewhat resembling the closing of the jaws of a cat on an unwary mouse, which she has captured by a similar process of strategy.

Notwithstanding their stupid, uninteresting appearance, as we behold them lying about on the bottom of the sea, many a covetous man would barter the everlasting salvation of his soul to possess the knowledge which some of their souls possess of the science of manufacturing pearls; and what would not their expert art in the mingling of the simple elements of color for the adornment of the interior of their pearl habitations be worth to the devotee of the art of painting! Unappreciable as the exterior of their habitation may appear, yet no king ever dwelt in a palace composed of such gem-like materials or adorned with paintings which could bear any comparison with the brilliant colors and exquisite harmony of the changeable effect of the shading or the burnished beauty of the interior of the tenement which they have built for their dwelling-place by the intuitive capacity which the Creator bestowed upon them in the harmonious mingling of equivalent proportions of sundry mites of the forces of animal growth from its department of nature's elements, whereby the meter of this second key-note on the D line of the natural staff in the harmony of intelligent enjoyment is made to chord with all the other parts of the entire piece.

If man is more intelligent than they, still he is proud of the adornment of his person with the pearls which can only be procured from their laboratory, and his parlors with the paintings which their pallets alone can furnish, and the dainty morsels of pastry prepared by their culinary art are among the choicest found upon kings' tables; so that as unprepossessing as they appear in their muddy laboratories, still their souls are at work within their limited circles, building, building, building, in the most expert manner, till they have completed a structure which no art of man can imitate, executed a painting which throws his best efforts into an obscure shadow, and perfected a gem which he is proud to wear. And the simple elements, both of the finite spirit of intelligence and force of animal growth, which by an equivalent combination in the composition of their souls, gives them their perfect capacity to perfom their little work, are drawn from the same departments of spiritual elements and force of animal growth from which the simple elements of the soul of man are taken, and he owes his physical, physiological, phrenological, and mental superiority over an oyster to an equivalent combination of the same simple elements of intellectual spirit and force of growth in the composition of the harmony of the intelligence of his soul that was used in the composition of the soul of an oyster, as surely as the most thrilling church anthem owes its inspiring effect to an equivalent combination of the same seven representatives of simple sounds which are used in composing the most discordant doggerel; and if man would use his extraordinary capacity with the same fidelity that an oyster does to the accomplishment of his own welfare and the general good of his species, how much wrong, oppression, and suffering would be stricken out of the calendar of crime.

The key-note of this second part in the grand scale of intelligent harmony is on D natural, and consists of an infinite variety of semi-tones, which are flatted and sharped into two plans of structure, one of which is called radiates, and the other mollusks.

The structure of the former (says Professor Agassiz, in his lecture at San Francisco, September 25, 1872,) is on the plan of a "single central cavity around which the organs are." This cavity is a sack, from which all parts of the body diverge, and into which they all open. These animals are all aquatic, and though the group has some fresh-water representatives, they are chiefly marine; from pole to pole, in all latitudes, the ocean swarms with them. This type is divided into three groups, namely: polyps, scalophs, and echinoderms.

"The variety of these forms is infinite; their beauty and complication of details are inexhaustible. Examine the internal arrangement of any polyp, be it sea-anemone, coral, or sea-fan, any of the countless and infinitely diverse acalephans, or any of the echinoderms, whether star-fish, sea-urchin, or trepang, and the secret of their structure is always the same. The idea of radiation underlies them all, and their typical plan may be described as a central digestive cavity, around which the various systems of organs are uniformly arranged.

"The graduation of rank in this division—that is, the comparative structural superiority or inferiority of the animals—is determined simply by a higher specialization of parts in some members of the group than others. In some there is an endless repetition of indistinct parts, their increase limited only by the life of the animal; while in others a fixed number of parts and clear differentiation of organs and of functional actions indicates a higher manifestation of the same structural idea.

"Next come the mollusks. The plan of structure is perhaps less easy to define in this than any other group, on account of the great power of contraction and expansion in all the animals belonging to it. The type is represented, like the radiates, by three classes, namely: the acephals, to which belong all oysters, clams, mussels, and the like; the gasteropods, under which come all snails, slugs, periwinkles, corals, &c.; and the cephalopods, among which are included the nautilus, the squids, and the cuttle-fish, of which your devilfish is so remarkable an example.

"In this group the idea of laterality, or a disposition of parts on two sides of the body, though with a tendency to compactness, which makes it less perceptible in the higher animals, is embodied. We have seen that in the radiates there is an equal disposition of parts, without reference to front or back, to left or right. In the mollusks, on the contrary, there is an unmistakable arrangement of parts with reference to the two extremities and the two sides of the body; and if I had time I could show you that the structural idea of this group is as unvarying as the radiates, and that all mollusks, whether bivalve, univalve, or of that singular group to which the cuttle-fish, squid, and nautilus belong, and in which the shell is sometimes hardly perceptible and at other times wholly absent, are constructed upon one and the same plan."

Thus, Professor Agassiz, who has made the form and nature of these and other animals the study of his life, testifies to the regularity of the structural type of the animals in this scale, and the constancy with which each specific soul follows after the spiritual pattern of its prototype, and has continuously repeated the typical structure thereof through all time, from the time that God composed the spiritual pattern up to the present time. As there is no evidence of any departure from a constant adherence to the rule in any instance, we are therefore bound to accept the axiom that like father like son, the axiom once an oyster always an oyster, and that the status of a cuttlefish, from the moment of its conception, is as immutably fixed for the purpose of earthly copulation and eternal spiritual enjoyment as that of a man is in his higher sphere, or the earth is in its orbit, by the equivalent spiritual elements and elements of force of growth selected from the second grades in their departments, which were combined in the soul of the prototype.

CHAPTER XII.

THIRD GRADE.

PISCATORY AND SNACAN FORCES OF ANIMAL GROWTH.

The key-note of this third grade of intellectual harmony is on the second line of the staff, assigned to the baritone, which is flattened into an embracement of all the vertebrated inhabitants of the sea, and sharped into a comprehension of all the reptile species.

In this department the simple elements of finite spirit and corresponding mites of animal growth are as variously arranged and combined into a stereotyped soul as there are species in both the flat and sharp scale, each of which has been increasing its specific family and populating the planets in the starry heavens, prepared for the eternal habitation of the beatific souls both of fish and reptile, where their everlasting little cup of fish and reptile joy is just as completely filled with joy and gladness as man is in his angelic sphere. And they fill their sphere in the measure of God's glory with more stable devotion than man does; for they have never rebelled against his moral government, nor cast his statutes aside, and set up instead thereof the statutes of reptiles or piscatory sovereignty, by which to work out robbery among their fellows.

Does anybody wonder why God made reptiles that sting and serpents that kill with venomed fang?

We answer: for the reason that, when he undertook the task of developing the resources of the department of nature embraced in the elementary atoms of finite spirit and mites of force of animal growth, his purpose was to compose every harmonious strain of intelligence that would chord with all parts of the tune of intellectual harmony and fill every sphere of enjoyment that a change in the specific arrangement of those seven simple elements of finite spirit could produce; which necessarily produced every specific form of body that could be produced by a change in the arrangement of the seven simple elements of force of animal growth. Each change in the arrangement of atoms composing the spirit of the soul of the prototype of a species of animal necessarily involves a corresponding change in the arrangement of the mites of force of animal growth; and a corresponding change in the form of the body and nature of materials composing it necessarily follows, whereby the degree of intelligence of the species is determined, actions prescribed, and mood of enjoyment everlastingly fixed to all the succeeding members of its race. Consequently, if God had neglected to make every change that could be made, and produce every specific degree of intelligence, form of body, and mood of enjoyment which could be made to chord with the universal harmony of the animal concert, by repeating itself in the exact pattern of the prototype in the lower grades as well as the higher, his work would be incomplete and his title to omniscience forfeited. Hence we see the reason why no new species can be formed by amalgamation. Every concordant species which any possible change in the arrangement and construction of simple elements could produce has been composed and was completed in the archetype of man; so that since his creation no new species ever has, ever will, or ever can be produced. Even God himself could not produce a new species without impeaching his former omniscience; for by ending his work in that department his omniscience was pledged that the changes were exhausted. Therefore if a new species could be produced, he would stand convicted of coming short of the end at which he aimed, which would bring him down from the infinite to the finite capacity.

Suppose a musical artist should contract for a stipulated compensation to exhaust the changes in arranging the scale of eight notes, and compose every tune and concordant part to each tune which a combination of all the harmonious changes could produce, and after having completed his work presents his book for acceptance and payment therefor;

whereupon the contractor submits the book to a committee of experts, who succeed in composing a tune, or concordant part of a tune, not contained in his book: he would lose not only his compensation, but his reputation of ability to complete would be irredeemably lost for having once said it was finished and come short of it: nobody would have any confidence in his ability to complete the task thereafter.

In order to complete that work, he would have to compose every doggerel song, childish glee, lover's serenade, political sonnet, national air, opera, sacred tune, doxology, and anthem possible to be composed therefrom. So also in composing the harmony of intelligence in the animal kingdom: when God said "it is finished," if another species of animalcule, shell-fish, worm or insect, fish or reptile, could be produced, his infinitude would be ended, and his capacity reduced to a finite sphere, and one sphere of intelligence, form of body, and mood of enjoyment would be vacant, and one star in the heavens without inhabitants.

God made a planet for the everlasting habitation of the beatific souls of the whole race of each species, before composing the archetype soul thereof to populate its own planet by copulation in this mundane sphere of gases, fluids, and gelatinous formations, which is simply the nursery from which the planets are populated.

Each species, looking only to the increase of its

own race, gives no heed to the welfare of the weaker races, and many carniverous races feed with impunity on all that they can master except their own. Here in this copulating sphere each race is a natural enemy to every other; therefore, the reason why many reptiles were armed with deadly weapons is very obvious. They have neither fins, arms, legs, nor wings, with which to escape from their enemies, which are legion. Their best effort at locomotion is but a slow crawling upon the belly; so that without some defence they would be as soon gobbled up as an oyster without its shell. Consequently, in order to secure the population of their planet, and fill their measure of harmonious joy in the infinite concert of animal life, God armed a sufficient number of them with venomous fangs to fill every species of animal with fear of anything in the form of a reptile, so that the few which are armed give protection to all the group, as a small, well-armed, and disciplined army gives protection to a great manufacturing nation.

Those armed reptiles are most honorable in the use of their armor, for they always give due warning before they strike, unless the attack is so sudden that they have no time to spring their rattle. Who ever heard the hiss of a serpent, appalling rattle of a snake, or felt the almost irresistible charm of its vindictive eye, gleaming over a pair of distended jaws bristling with venomous fangs, between which the tongue, like forked lightning, darts in and out of the elevated head, the very acme of self-reliant power, as it glides on to the capture of its victim with an imperceptible movement, while his rattle plays a fascinating concert, and did not feel that every movement was instigated by an immortal spirit, and the fascinating charm was the effect of mind over mind? Who ever saw a demoralized bird leave its perch on a tree, and in bewildering curves draw nearer and nearer to a point, till it finally fluttered straight into the distended jaws of the monster, and did not believe that it was the effect of a strong spiritual magnetizer over a mesmerized subject; or the exceeding care which every animal manifests in avoiding its treachery, whether discovered by smell or rattle, and did not believe that this fear was produced by a knowledge of the deadly effect of the venom of its bite?

If there is a spirit in such reptile, then the structure contains a soul, the celestial substance and spiritual forces of which can neither be decomposed, annihilated, or changed in form to the breadth of a hair in any material dimension, and therefore must forever remain a self-reliant fascinating reptile, though it never captures a bird except a sense of hunger demands it, or attacks or even strikes an enemy, except in self-defence or the defence of its young.

Fishes have finny propellers, exactly adapted to the watery element in which they live, and some even take to themselves wings, leave the water, and skim through the air when in great peril, till their enemy has lost the scent.

Whales bring forth their young from a fœtus, in a living state, and suckle them like bipeds; and each species of fish is as constant to the spiritual pattern of the soul of its prototype as the race of man is to his; and each race has a mood of enjoyment as peculiarly its own and differing as widely from any other as it differs in form of body and texture of flesh, and is a natural enemy to every other; and there is about the same number of carniverous species that feed upon others as there is among other animals.

Who ever saw his fellow gobbled up by a shark, that did not see in it a contest between antagonistic minds in an element where the will of the shark was stronger than man's, and the soul of the shark had an easy victory over the soul of the man. Yea, verily, every fish which is produced by copulation from the time of its conception is a living soul, whose degree of intelligence, mood of enjoyment, and physical appearance is a concordant pattern of the archetype of its species, and its status in the tune of intellectual harmony everlastingly determined.

As we have seen that the spiritual pattern of every species of animal which can harmoniously reproduce their prototype has been produced; therefore, every cross that is made between the species produces a discord in the universal harmony of the animal kingdom, in which the hybrid is a despised abortion, and in which there is a perpetual warfare between its own members to such an extent that many of them will not breed at all, as in the case of mules, and in those that will breed, the soul of the sire works it up or down to its own specific standard at the seventh generation, as in the case of Angora with Spanish goats, or man with negroes or Indians.

Every race of animals is at enmity with a hybrid till it is restored to its type, and its own members are at enmity between themselves to such an extent, that its best estate of existence is only a living death. Hence the divine commandment, Thou shalt not amalgamate thyself, nor let thy beast; and death is the penalty for any man or woman who has sexual commerce with any other species of animal—death to both the guilty parties, and for causing other animals to amalgamate: both male and female, and offspring, if born before discovery, are to be put to death, and the loss of the carcasses to the owner; and fivefold their value against the guilty party if he did not own them.

No two species ever did amalgamate except they were excited to it by the irresistible influence of man. Different species of fish will feed upon each other, but they will not amalgamate; different species of reptiles will feed upon each other, but they will not amalgamate. The whole burden of this sin lies on man's shoulders: he alone has violated this most sacred law of his nature, and he alone has power to cause other animals to perpetrate the same sin.

Each species of fish cleaves to its own race for sexual intercourse, and they are as constant to their archetype in propagation as a tree is to bring forth seed after its kind, or the seed of a tree is to repeat its specific type. Minnows, chubs, dace, trout, shad, pike, salmon, mullets, pickerel, white-fish, suckers, bass, catfish, herring, cod, and the thousands of other species grow on from generation to generation. One generation follows another, on to the stage of action through the door of copulation and through the dark curtain of death to their celestial homes, with the same constancy to the spiritual pattern of the soul of their prototype that men pass on and off the stage of action in their successive generations. The golden specks once formed on the skin of a trout will never cease to glisten; neither will his intelligence ever depart from him. The soul of each trout, from the day of its conception, is as permanent a fixture in the everlasting concert of intelligent enjoyment, love to God and its fellows, as man's is, or as the earth is a fixture in the planetary system, for the reason that the spiritual pattern is composed by an equivalent arrangement of the same simple elements of spirit and forces of animal growth—

the same elements of substance, celestial and terrestrial, which, being differently arranged, compose the soul of a man; and both are arranged by the same infinite intelligence, combined and constituted into a living soul by the force of the same holy spirit. Therefore, if the one is immortal, the other must be; if the sagacity and intelligence of one are the work of reason, so are they in the other; if the purposes of God are irrevocable, then both are immortal.

CHAPTER XIII.

FOURTH GRADE.

EMBRACING THE FORCES OF ANIMAL GROWTH EMPLOYED IN THE PRODUCTION OF WORMS AND INSECTS, CALLED ARTICULATES.

The key-note to the intellectual harmony of this fourth grade of finite intelligence, called articulates, which puts it in chord with all the other parts in the great piece of universal intelligence, is on F, the second space of the staff in the natural scale; and as we have seen that scale G constitutes the baritone part of the piece, so this evidently fills the alto grade of the staff. If the force of animal growth employed in this scale exhibits less control over heavy material, yet it certainly performs its work after the spiritual pattern with much greater activity, structure more graceful, and some of the species actually sing their parts in thrilling chorus with the universal concert, while others light up the gloomy nights of autumn with steady glows or brilliant flashes of light of their own manufacture, and some of them enter into a well-contested rivalry with the shell-fish in the art of mingling colors and pleasing illustrations, which if less polished and durable than the pearl manufactories thereof, are yet more brilliant and fanciful, and the solid well-filled cards of honeycomb with which the laboratory of the industrious little bee is yearly filled for the sustenance of man is of much more general utility than a pearl necklace. Thus each social scale plays its harmonious part in the universal concert of intellectual souls with the uniform approbation of their Creator and equal enjoyment to themselves, which when individually played out in this mundane sphere of existence, their liberated souls are taken up to their eternal home in the starry heavens, which God prepared for them ere the spiritual pattern of the prototype was composed, where they continually perform their parts in the anthems of praise and thanksgiving to God, who composed every part of the piece from the same scale and cares for all alike.

Having introduced Professor Agassiz's testimony in the above second grade, we will also give an extract from the same lecture, delivered in San Francisco, September 25, 1872, on this third grade, viz:

"Next we have the articulates—all the hosts of worms, crustacea, and insects. The structural idea is that of a cylinder, divided by joints into movable rings; and whether we have the body of a worm articulated from end to end in equal sections, or whether some of these sections are soldered together to form a front part of the body distinct from the posterior portion, in which the sections are left free and movable, (as in the lobster and crab,) or whether the sections are broken into three parts, to form head, chest, and hind body, (as in the insects,) the structural idea is invariable. The body of worms, crustacea, or insect is always a hollow cylinder, containing a variety of organs and divided by articulation.''

The professor, in this lecture, gives a concise scientific description of the structural plan on which both this third and the above second grade in the animal kingdom are produced, to which we will add, that each specimen in all three of the above grades has a heart, which is the seat of the force of animal growth, from which the work of constructing the body and repairing the system from day to day is carried on; also a head, containing the machinery of the mind, which is the terminus of the nervous system, where the spirit of intelligence has its seat, and is the prime cause of every sensation felt in the nervous system, produces every muscular movement, guides and directs all the goings out and comings in of the body, and produces every thought and intelligent idea of the soul, without which the structure, when completed by the force of growth operating from the heart, would be as inert and senseless as a vegetable. ' The force of animal growth having completed the structure, has no more power to quicken it into the condition of a sensational living soul than the laws of crystallization have to produce vegetation, or gravity has to produce an earthquake. Throughout the entire system of organic bodies each force is assigned a definite action, and operates from a stand-point of its own to a given end, and is as constant in producing a uniform effect as the sun is in illuminating the universe.

The force of animal growth constructs the body after the spiritual pattern, embracing the vital organs, and strung with a nervous system operating from the heart, while its superintending co-worker, the finite spirit of animal intelligence, has its stand-point of operation in the brain, and quickens the structure into the sensational sphere of a living soul, capable of thinking, reasoning, willing, and acting from intuition, which is as surely the cause of each structure and executive volition thereof, first in the grade of animalcules, second of mollusks and radiates, and third of articulates, as it is in the race of man; and the distinction between the grades of intelligence and the "comparative structural superiority or inferiority" is determined simply by a change in the arrangement and combination of the same simple elements of finite spirit and force of animal growth in composing the harmony of the spiritual type of the species, as one arrangement of the same scale of eight representatives of simple sound composes a doggerel song, and another a heaven-inspiring anthem giving praise to God; or as one arrangement of the alphabet composes a treatise on scandal, another on medicine, another on ethics, another on metaphysics, and another on theology, all of which are language, communicating ideas from mind to mind, each representing a certain train of thought inscribed by the author in words, syllables, and sentences, composed, by a series of varying arrangements, of the twenty-six letters of the alphabet, only seven of which represent simple sounds. So also all the animals in the grades of animalculæ, shell-fish, worms, and insects are intelligent souls, capable of originating ideas and communicating the same to each other, each species of which represents a train of thought in the mind of the Creator, inscribed by the varied combination of seven simple elements of finite spirit from the same department.

Who ever viewed the orderly step with which the multitude of legs on a centiped take up their line of march, and the martial dignity with which every turn of the body is made, the defiant regularity of march, and the flashing eye under provocation, without seeing in every movement the dictation of an intelligent spirit, and a confiding trust in its own discipline and the venom of the fangs with which it is armed?

Who ever teased a scorpion into a war passion, and witnessed its strategic movements, springs from side to side in evading a blow, sudden feint of retreat and more sudden spring of return to the attack, with its venomous dagger flourished high in air, ready to strike a fatal blow in every direction, and did not see in it the courage and training of a gladiator?

Who ever viewed the serene light of the delicate little glow-worm's own manufacture, its active somersaults, and bright eye, that did not believe that little spirit which looks out through the eye causes the movement and manufactures the light?

Is there no intelligence in the mathematical plan of a spider's web, and the precision with which the lines are laid to and from the exact points which give them the greatest strength? Who ever was struck with the dagger of a hornet, that did not feel that it was driven by a vindictive spirit in avenging some real or imagined wrong? Who ever partook of the delicious pastry from the laboratory of a colony of bees, that could say he believed it was prepared without a purpose? Who ever listened to the nightly concerts of the crickets under the hearth, katydids among the garden shrubbery, and locusts in the park, and did not hear anthems of praise and thanksgiving to him who created us all?

CHAPTER XIV.

FIFTH GRADE.

FOWLS.

The forces of animal growth, as arranged and combined from the simple elements thereof for the production of fowls in this fifth group in the animal kingdom, are certainly experts in the mingling of the simple elements of color, and stand at the head of the list of illustrating artists in the animal kingdom, and are scarcely excelled by the floral artist in the growth of flowers.

If the terrestrial part of the plumage in which the pictures are indelibly impressed is less durable than the oyster's pearl, the celestial substance thereof is indissoluble, and preserves the illustration forever. If the terrestrial plumage is less transparent in material or polished on its surface than the pearl, the celestial is clearer, and the colors are reflected more brightly to the spiritual eye.

This mingling of the elements of colors in bodies that grow is the work of the force of growth in the animal as well as the vegetable kingdom. The intelligent spirit of the mind has nothing to do with the coloring of its flesh, skin, hair, or plumage. The entire work of embellishment of natural bodies is performed by the force of growth, acting from the heart, and independent of any mental dictation from the head, yet in perfect harmony with the spiritual pattern of the soul; so that, whatever the design or coloring of the picture may be,—whether of hair, skin, plumage, or shell,—it is the glory of the mental sensation of the soul and pride of the intellect, which would feel humiliated and disgraced by the slightest change in a shade of its natural color. The force of growth, operating from the heart, produces the body in the specific likeness of its spiritual prototype—a physical, physiological, and phrenological development of the immutable purposes of God in composing the species of which it is a propagated specimen, and embellishes it with the destined colors, which, though varying slightly in design, are of the same specific pattern, which no power but him who formed them can change or alter, nor sin of amalgamation change for more than seven generations. All specific concordances in the propagation of animal bodies by regular specific copulation are an infinite repetition of an omnipotently designed pattern. All amalgamating discords produced by irregular copulation are only a finite disarrangement of an infinite pattern, limited to the seventh generation at the furthest. The disarrangement is only in the terrestrial matter of the body in which the soul is molded—

the terrestrial mold in which a celestial jewel is cast—the terrestrial matter in which the substance of an immortal soul is propagated. The discord is produced by the lodgment of the fœtus in a foreign womb, where it is compelled to draw its supplies from an unnatural reservoir. No amount of amalgamation can affect the form of the spiritual pattern or component substance of the celestial body. Hence the hybrids produced by any cross between different races which will breed together work back to the genuine type of the sire at the end of the seventh generation, and that is the end of the discord.

The key-note of this fifth part in the universal tune of harmonious inte'ligence is on the G line of the natural scale appropriated to the contralto, and is actually sung by the different species on every variety of flats and sharps, high and low keys, yet always in perfect chord with the seven parts of the entire piece representing the complete intelligence of the animal kingdom.

What doxologies of praise to God are poured forth from those little throats; what anthems of thanksgiving ascend to heaven from those concerts of feathered songsters; what loving warblings are sung in the pairing season, and what soothing chords of lullaby are addressed to the tender young by the mother nurse while pa is capturing the nourishing worm or gathering cereals to develop them into a new choir of feathered songsters, who, having sung a few songs for our entertainment, and added a few broods to the number of their species, drop the terrestrial dross of their gay plumage, and arrayed in the celestial substance thereof, refined into its gorgeous celestial brilliancy, flit away to their eternal homes, join the throng of their ancestors in their allotted sphere in the starry heavens, from which an everlasting song of praise goes up to him who composed both them and us of the same elements, that each should fill a sphere in the halo of his glory, and the intellectual cup of each be filled with overflowing joy.

Who ever heard the song of a bird that did not want it to live forever and warble off its overflowing joy in songs of praise to its Creator. And are we to suppose that God would spend his energy to compose these choirs of sweet songsters one day, to strike them out of existence the next? No, verily, the purposes of God are without recall, and the spiritual patterns of his designing are forever the same: once a linnet always a linnet. And the lark, in the ascension of its morning devotion, only essays the final ascension which its liberated soul takes when it drops the dross of the mold in which it was cast.

If we are edified by the melody of their concert who are strangers to them, how much more is God glorified when he hears the pleasing effect of his own composition: and Christ said that not a sparrow falls to the ground without his notice.

If the force of animal growth, as combined for the production of this group of species in the animal kingdom, is an expert in the mingling of colors for their adornment, the spirit of intelligence is an equal expert in the melody of song, for it has composed strains, yea, verily, and sung them too, so perfect in time, tune, and inspired melody, that man hath never yet attained to its equal or fair imitation.

The concord with which some of the groups in the animal kingdom harmonize in their intelligent acts is beautifully described by Cowper in the following extract from a poem:

> "A nightingale, that all day long
> Had cheered the village with its song,
> Nor yet at eve his note suspended,
> Nor yet when eventide was ended,
> Began to feel, as well he might,
> The keen demands of appetite:
> When, looking eagerly around,
> He spied far off upon the ground
> A something shining in the dark,
> And knew the glow-worm by its spark;
> So, stooping down from northern top,
> He thought to put him in his crop.
> The worm, aware of his intent,
> Harangued him thus right eloquent:
> 'Did you admire my lamp,' quoth he,
> 'As much as I your minstrelsy,
> You would abhor to do me wrong,
> As much as I to spoil your song,
> For 't was the self-same power divine
> Taught you to sing and me to shine:
> That you with music, I with light,
> Might beautify and cheer the night!'"

SIMPLE ELEMENTS OF ANIMAL GROWTH. 163

Thus all act out their own intelligent parts in the universal opera of intellectual harmony for their own amusement, the entertainment of each other, and the glory of God, who composed them all.

What though the nightingale did put that glow-worm in his crop, which no doubt he did, notwithstanding his admiration of the glow and the harangue; it only quenched the terrestrial spark, while it liberated the soul of the worm from the dross of the mold in which it was propagated, which, adorned with its celestial glow, was taken up by its Creator, and joined with its ancestors in their everlasting home in the heavens, where they will glow on forever to their own satisfaction, entertainment of the beatific souls of all the animal kingdom, and the glory of the Omniscient Composer. And it is just possible that some greedy hawk parted the soul of the nightingale from its propagating dross, by gobbling it up ere the close of the succeeding day, when his soul flew off with its celestial song to its eternal home in the heavens, and again joins music to light with his old comrade in the everlasting concert of intellectual opera.

In this group, as in all the other seven, there is an infinite variety of species, which never cross or amalgamate, however much they may resemble each other, though associating in the most friendly manner in the same domestic barn-yards, wooded parks, and limited groves, where the males and females of each race pair off in holy wedlock, copulate, and rear their little families, as constant to the spiritual pattern of the soul of the prototype of their species as the sun is in the immutable light of its illuminations.

The lover of nature can find a large open book of instruction in every well-supplied colony of domestic fowls, in which those intelligent pedagogues are ready to impart a vast amount of information, if he will be taught, which will strongly corroborate the truths herein set forth.

Such a student was Addison, and what he learned in that school we will let him relate in his own way by the following quotation from one of his own speculations: (Addison's Works, by Richard Hurd, volume 2, page 457:)

"My friend Sir Roger is very often merry with me upon my passing so much of my time among his poultry. He has caught me twice or thrice looking after a bird's-nest, and several times siting an hour or two together near a hen and chickens. He tells me he believes I am personally acquainted with every fowl about his house; calls such a particular cock my favorite, and frequently complains that his ducks and geese have more of my company than himself.

"I must confess I am infinitely delighted with those speculations of nature which are to be made in a country life; and as my reading has very much lain among books of natural history, I cannot forbear recollecting upon this occasion the several remarks which I have met with in authors, and comparing them with what falls under my own observation: the arguments for Providence drawn from the natural history of animals being in my opinion demonstrative.

"The make of every kind of animal is different from that of every other kind; and yet there is not the least turn or twist in the fibers of any one which does not render them more proper for that particular way of life than any other cast or texture of them would have done.

"The most violent appetites in all creatures are lust and hunger: the first is a perpetual call upon them to propagate their kind; the latter, to preserve themselves.

"It is astonishing to consider the different degrees of care that descend from the parent to the young, so far as is absolutely necessary for the leaving a posterity. Some creatures cast their eggs as chance directs them and think of them no further, as insects and several kinds of fish; others, of a nicer frame, find out proper beds to deposit them in, and there leave them, as the serpent, the crocodile, and ostrich; others hatch their eggs and tend the birth till it is able to shift for itself.

"What can we call the principle which directs every different kind of bird to observe a particular plan in the structure of its nest, and directs all of the same species to work after the same model? It cannot be imitation; for though you hatch a crow under a hen, and never let it see any of the works of its own kind, the nest it makes shall be the same, to the laying of a stick, with all the other nests of the same species." * * *

"Is it not remarkable that the same temperature of weather which raises this genial warmth in animals should cover the trees with leaves and the fields with grass for their security and concealment, and produce such infinite swarms of insects for the support and sustenance of their respective broods? Is it not wonderful that the love of the parent should be so violent while it lasts, and that it should last no longer than is necessary for the preservation of the young." * * *

"With what caution does the hen provide herself a nest in places unfrequented and free from noise and disturbance! When she has laid her eggs in such a manner that she can cover them, what care does she take in turning them frequently, that all parts may partake of the vital warmth! When she leaves them to provide for her necessary sustenance, how punctually does she return before they have time to cool and become incapable of producing an animal. In the summer you see her giving herself great freedoms and quitting her care for above two hours together, but in winter, when the rigor of the season would chill the principle of life and destroy the young one, she grows more assiduous in her attendance and stays away but half the time. When the birth approaches, with how much nicety and attention does she help the chick to break its prison! Not to take notice of her covering it from the injuries of the weather, providing it proper nourishment, and teaching it to help itself, nor to mention her forsaking the nest if after the usual time of reckoning the young one does not make its appearance. A chemical operation could not be followed with greater art or diligence than is seen in the hatching of a chick; though there are many other birds that show an infinitely greater sagacity in all the forementioned particulars." * * *

"As I was walking this morning in the great yard that belongs to my friend's country house, I was wonderfully pleased to see the different workings of instinct in a hen followed by a brood of ducks. The young, upon the sight of a pond, immediately ran into it, while the stepmother, with all imaginable anxiety, hovered about the borders of it to call them out of an element that appeared to her so dangerous and destructive. As the different principles which acted in those different animals cannot be termed reason, so when we call it instinct, we mean something we have no knowledge of. To me, as I hinted in my last paper, it seems the immediate direction of

Providence, and such an operation of the Supreme Being as that which determines all the portions of matter to their proper centers."

It is strange that Mr. Addison did not discover that the acts of those fowls were as much the effect of a spirit of intelligence operating upon their brains as ours are, and that, as far as their intuitive capacity extended, their actions were directed by a will, produced by reasoning from cause to effect, and a judgment of the effect certain to follow a cause, which affected their interest, that is wonderfully correct. But as he was not able to discover this truth, probably on account of its simplicity, nor able to distinguish between our reason and their intuition, he fell back upon the heathen idea, *Deus est brutum*—God himself is the soul of beasts—which for so many centuries filled Europe and Africa with gross moral darkness, and caused the sacrificing of thousands of innocent people on the burning altars of idolatry to appease the wrath of devouring beastly deities.

Christianity commenced a war of extermination on that demoralizing pagan theory of a devouring beastly deity about eighteen hundred and seventy-three years ago, since which time it has been steadily dispelling that miasmal fog of moral death, and opening up a highway for the progress of scientific research into the great truths which were so long hid from men by heathen mythology and religious bigotry. Progressive research into heavenly truths, which Christ compared to a grain of mustard seed producing a tree, a spoonful of yeast in a barrel of flour leavening the whole lump, and many other progressive things, has finally exposed the absurdity of a deity having infinite parts without a head, and has revealed in its stead a God of infinite parts, with a self-constructed omnipotent person, crowned with an omniscient head to devise and a force of holy spirit to execute through an omnipresent system of nervation, acting from a common center, in composing the universe, as the rays of the sun do in illumination of it, who has composed all the specific degrees of intelligence in the animal kingdom by changing the arrangement of seven simple elements of finite spirit in affinity with a corresponding change in the seven simple elements of the force of animal growth; consequently, each change in the combination of spiritual elements causes a corresponding change in the combination of the elements of the force of growth, which changes in some specific degree the physical, physiological, and phrenological form of the body, texture of flesh, turn and strength of muscle, tone of the nervous system, and degree of intelligence, which necessarily change the intuitive capacity for originating ideas, and limit the reasoning capacity to a circle of greater or less extent—the boundaries of which are determined by the field containing the resources for supplying its own wants.

Each change in the texture of flesh requires a corresponding change in the quality of food for its replenishment, and in the mode of obtaining it. Each specific change in degree of intelligence changes the mood of enjoyment. Hence each species of animal by intuition adopts a different habitation, and works out its destiny of life in a mood differing from every other in about the same degree that it differs in personal appearance.

Ducks and other water-fowl have webbed feet, which serve as propelling paddles in the water; hens have long strong toes and clumped nails, with which to scratch for worms and seeds on dry land; hawks and other birds of prey have strong claws and sharp nails, with which to seize their victims, and strong wedge-pointed beaks, with which to strike a deadly blow and tear the carcass to pieces.

In domestic fowl the facilities of studying the adaptability of structure to mood of enjoyment and manner of working out the destiny of life are much greater than in others. With what wonderful constancy they adhere to the spiritual pattern of their prototype. Change the hen's eggs for those of a duck, and the biddy nurse will not only hatch them, but she will nurse them with the same assiduity that she would her own brood; and notwithstanding she despises the fraud, still she remains true to the trust reposed in her, covers them with her wings, and scratches for them to the extent of her ability till they are capable of providing for themselves, and that is the end of the fraternal relation between them. Amalgamation never follows this motherly care and infantile dependence. The period requiring protection ended, the mother and adopted ducklings each take up their specific work of destiny, and go on with the propagation of a holy seed in harmony with the spiritual pattern of their race. Never in any case do they mate with another species.

The constancy with which the spiritual pattern in which God composed the soul of the patriarchal pair of each species, by an equivalent combination of the seven simple elements of finite spirit, is easily observed and clearly demonstrated by fowls bred in the same yard and under the same force of circumstances. Here the axiom is clearly and strikingly demonstrated. But it is no more marked and rigorously observed by them than it is by every other species throughout the animal kingdom below the grade of man. Even the thoroughbred Indian never did, and never will, attempt to amalgamate with negroes, apes, monkeys, or even a woman or other female not of his race, which was demonstrated in the early settlement of this continent: for in the early Indian wars there is no record that they ever attempted to amalgamate with captive women. Whatever else they did, they always maintained the purity of their race. All the half-breed Indians and mulattoes are the result of a cross between white men and female Indians or negresses.

Man is the only animal that ever deteriorated his own soul or vitiated the spiritual pattern of his species by voluntary amalgamation. In every other species of animal, from the Indian to the animalcule, the constancy with which each species confine their matrimonial alliances within the limits of their own race, and repeat the spiritual pattern of their prototype, was beautifully expressed by the poet when he sung—

"Like loves like, and love likes love;
Eagle mates with eagle, and dove seeks dove."

Thus the animal kingdom instinctively classifies itself into species, and demonstrates the manner in which the seven simple elements of the force of animal growth were arranged in the spiritual pattern of the first pair by the physical, physiological, and phrenological development of the structure, and the manner in which the seven simple elements of finite spirit were arranged in the construction of the spiritual type of the species by the intuitive ideas of the race, their specific mood of enjoyment, habits, and manner of working out the destiny of life.

There is no axiom within the comprehension of man better established than that the families of each species build their habitations, burrows, nests, and marine copulating beds on a uniform plan of structure. Hence, wherever a family of

Indians is found, whether in Asia, Europe, or America, their wigwams are built on a uniform plan of structure; and all the mud huts built by the families of negroes are as uniform in structure as the nests of crows, martins, or hanging birds; and whatever form of architecture or plan of structure any species of animal build, we should know that their habitations are by their own intuition exactly adapted to their mood; and to compel or induce them to live in any other is forcing or seducing them into a state of barbarism, living in a condition contrary to nature. Men who have the control and management of animals should always watch their moods, and furnish them with habitations as near as possible like those which they construct or choose for themselves, and supply them with the food of their own choice. Neither should different species ever be herded in the same fold, always bearing in mind that such change in the arrangement of the seven simple elements of spirit causes a similar change in the arrangement of the seven simple elements of force of growth, and a corresponding change in the arrangement of the seven simple elements of celestial substance and terrestrial matter which compose the specific body, and necessarily requires a change in the kind of food to reconstruct and keep it in repair. In all cases the owners of poultry should carefully observe the intuitive desire of the soul of the species as to quality of food, nesting places, and shelter from storms, and the very largest liberty should be allowed which is consonant with domestic order; and any oppression, starvation, or abuse of them should be promptly punished by the government. They are living souls, which God has composed by a different arrangement of the same simple elements of spirit, force, and substance of which man is composed, and are capable of enjoying a great deal of pleasure and of suffering intense pain, and he will bring their owners to judgment for any abuse of them.

CHAPTER XV.

SIXTH GRADE.

ELEMENTS OF FORCE OF ANIMAL GROWTH—QUADRUPEDS.

The mites of the force of animal growth, as arranged and combined into spiritual patterns for the propagation of the various races in this grade of the animal kingdom, are on a much larger plan and more substantial basis than either of the five lower grades.

As it is in all the foregoing groups, so it is in this: there is an infinite variety of species, between all of which there is some essential difference in structure, habits, and mood. Still the structural plan of the whole group is based on the quadruped idea—vertebrates, with four substantial legs, tipped with strong elastic feet, some of which are coronet shaped, some cloven, and some with several toes, shod with durable elastic soles, and bound together with strong ligaments or elastic webs.

The physical appearance of each species in this group exhibits the exterior of the spiritual pattern of the soul of its archetype, the nervous temperament demonstrates the tone of the nervous system on which it plays its own mood, and the phrenological development indicates the ideal and mental capacity of the spirit.

The arrangement of the simple elements of the force of growth in composing the soul of any species of animal is always combined in such perfect harmony with the arrangement and combination of the same simple elements of finite spirit, that the length, size, and shape of every bone, form of muscle, tension of nerve, vital organ, and phrenological bump are composed in perfect concord with the mental forces and capacity of the spirit; so that by studying the exterior of the structure, the spontaneous acts which demonstrate the intuitive idea as clearly as words could do, and the physical strength of bone and muscle as manifested by their movement, and natural propensities by their habitual moods, we can judge very correctly of the plan on which the simple elements of force and spirit were arranged in composing the spiritual pattern of the soul of the species. And by observing the kind of food which it voluntarily selects at the table of nature, we can judge tolerably well of the proportionate combinations of matter used in its specific structure, and thus be able to treat those over which we have control in a manner most conducive to their enjoyment, and consequently to their health, fattening condition, and physical and mental powers of endurance to perform good service.

In said investigations into the nature, wants,

and best mode of treating every species of animal over which we have control, it will be well to remember that each specific change in the arrangement of the seven simple elements of finite spirit, in composing the spiritual pattern of the species, causes a corresponding change in the arrangement of the seven simple elements of force of animal growth and the seven simple elements of substance and matter constituting the soul. The inevitable consequence of which is a corresponding change in the length, size, and shape of every bone, rounding and fitting of every joint, size, shape, and twist of every muscle, tension of every nerve, texture of flesh, size of veins and arteries, and a corresponding change in the sanguine fluid which circulates through them, which necessarily requires a corresponding change in the food that supplies the material of which the blood is composed for the construction and repair of the body.

It is true that a species of animal can be made to eke out a sickly existence on food which contains but very little of the elements required by their system, but it is a great waste of food, as well as a cause of much suffering to them; for they are compelled to eat enormous quantities of it in order to extract therefrom the very little nourishment which it contains for their system, which keeps their stomachs crowded and overloaded with foreign substances which are in a manner poison to them, but just the elements which some other ani-

12

mal requires for its proper nourishment, and which is thus destroyed by an unnatural digestion, is a source of irritation to the bowels, and is finally thrown off in piles of excrement, to be trampled under foot by the very animals that would have fattened upon it if it had been fed to them in its cereal condition, and is thus not only lost to the animal kingdom, but becomes a putrescent manufactory of unwholesome gases, engendering disease through the fold.

Much valuable time is also lost, force unnecessarily expended, and nutritious cereals destroyed by cooking and fermenting food for hogs and other animals, as is frequently done. The stomachs of all the species of this group are capacitated to digest their food in its natural cereal condition. Cooking and fermenting only create an unnatural appetite, whereby the stomach is overloaded and the contents hurried off through the intestines in an undigested condition, irritating the bowels and producing a feverish condition of the system, which render the meat soft, oily, and vapid. Potatoes fed raw to hogs will make about one-third more pork, which is much firmer and healthier for food than when cooked or fermented before feeding. It is true it will take a little more time to bring them up to the same condition, but the saving in food, expense of cooking, and quality of the meat doubly make up for the time: *i. e.*, grown hogs cannot be brought up from starved skeletons to a killing condition on raw potatoes in as short a time as they can on boiled and fermented mash; but if pigs are fed regularly with all they can eat, from the time they are born till they are eight or nine months old, they will be in better condition at the end of nine months than skeletons starved for sixteen months and fattened for three on one third less food, and the meat will be worth one quarter more per pound.

Feeding work horses or other stock with chopped or ground oats, barley, or other cereals is a waste of food and injurious to the animal. It is necessary that all the food taken into an animal's stomach should be thoroughly saturated with the saliva from the glands of the mouth before passing into the stomach to prepare it for digestion, which can only be done by thorough mastication between their own grinders. Ground food is swallowed without being properly saturated with the saliva from the glands, consequently it is hurried on through the system in an undigested state, operating as a cathartic, and prostrating the animal instead of strengthening it. Thus, what would have been nutritious strengthening food, if ground between the jaws of the animal, becomes prostrating physic when ground before feeding, and the result is a waste of food and prostration of energy in the animal it was designed to strengthen.

In this group of the animal kingdom, as well as with men, the mental condition has great influence over the physical forces. Therefore, in order to get the best use of servants, whether quadrupeds or bipeds, they must be kept in good humor, and supplied with food most congenial to their systems, so that no force be wasted in digesting unwholesome food, or lost by a physic-weakened condition of the system.

The key-note of the intellectual harmony of this group is on A of the natural scale, appropriated to the tenor part of the tune which comprehends the universal drama of the animal kingdom, and on the fifth line of the staff, and is flatted and sharped, (1) from a very small mouse to a grizzly bear; (2) from a pig to an elephant; (3) from a weasel to a royal tiger; (4) from a petit dog to a lion; (5) from an antelope to a buffalo; (6) from a goat to an ox; (7) from a pony to a draught-horse. All the species in this group perform their part in the drama by acting; there is not a musician in the group, but what they lack in musical talent they make up in force of action, many of which are of great utility in the work of civilization, under the judicious management of man.

Each species has a language precisely adapted to the understanding of its own race, made up partly of words vocally expressed, partly of gestures, and partly of a spiritual reading of each other's thoughts, by which they exchange ideas: and a man who is not capable of learning their language and understanding their requests, desires, and ailments does not know as much as they do, and never should be allowed to have any control or management of them.

Any of them learn our language readily; and if they had organs of speech which admitted of a proper pronunciation of words in our language, many of them would converse with more decorum than many of the bipeds of our own race; and with most of them the best of us could swap ideas with profit to ourselves.

Among those animals which are employed as servants in a variety of useful occupations the horse stands at the head of the list, as a kind, intelligent, faithful servant, which could not well be dispensed with in tilling the earth, gathering crops, transporting them to market, propelling our family carriages, and transporting army supplies, in all of which occupations they perform nineteen-twentieths of the labor which men would otherwise have to do; and when used at any one employment steadily, they soon learn to perform it skillfully, without whip or rein. Having been driven a few times over a road, drawing a loaded vehicle, they recollect every rough place, crook, and turn in it, and will work a load through it in a night so dark that the driver can render no assistance in guiding them; and if kindly treated, will perform their duties at the word of command just as honestly and faithfully as man can.

Many cases of extraordinary intelligence and

fidelity in the performance of certain kinds of service by them are published in natural history, and we will add one more to the list which came within our own knowledge.

While traveling through the gold mines of California, we visited a placer claim, situated about a quarter of a mile from the American river, from which the owners were hauling gravel to the river and washing out gold in sluice-boxes with water taken from the river. At the time we arrived at the claim they were loading a cart with gravel in a pit which had been sunk some four feet to uncover the strata of pay gravel. Hitched to the cart was a substantial, strongly-made pony, of the French stock, raised in Canada. When the cart was loaded, one of the men said to the pony, "That's a fair load for you; take it down," and away he went to the dump-pile, wheeled around, and backed up as near to the sluices as possible, and quietly waited till the man who was washing got the sluices in a condition to leave them, when he dumped the load. Away went the pony again to the pit, wheeled around, and backed into the place of loading, without any word, sign, or gesture from the men, and when the load was completed returned to the dump, and then back to the pit; and thus he went on till the day's work was ended, when we all went down with the last load to see the panning out of the day's work. As soon as the last load was dumped, the pony started off again, and we inquired if he was going back to the pit. They answered, "No; he is going to the stable this time, where his mistress, whom he brought across the plains on his back, will ungear him and put him in his nice stable, and pay him for his day's work with a lump of sugar, a feed of good well-cured hay, some nice barley, a few friendly pats about the head and neck, and he believed she kissed him sometimes, all of which he accepted with a friendly whinny, a kiss in return, a bow of thanks, and considered himself well paid for his day's work.

This kind of intelligence in a horse may seem wonderful to people who suppose that a horse is only a little spark of phosphorus done up in a bundle of lime, sulphur, saltpeter, and mucilage, which a scratch may touch off with a snap, a fiz, and a blue blaze with an unpleasant odor, and their spark of life is extinguished forever: only a bundle of animal instinct, which appears for a year or two in the form of a horse and then disappears, to make his next bow to us in the form of a toad or a snake. But to those who understand the plain simple truths of nature, it is no more wonderful than their own intelligence. In that honest, faithful, working horse they see only a different arrangement and combination of the same simple elements of spiritual force and substance in equivalent proportions to compose a horse and give him a certain degree of intelligence and reasoning capacity, which being differently combined in equivalent proportions to constitute a man, give him a higher degree of intelligence, a larger circle to his reasoning capacity. As one arrangement of the simple elements of sound signifies MAN and another signifies HORSE, the simple elements of sound pronounced in a certain manner signify horse: so also the simple elements of spirit, force, and substance, combined in a certain manner, compose an intelligent soul. The quality of sound is signification; the quality of soul, intelligence. Sound signifies the name and quality of a thing; soul expresses the sound and understands the signification. As the horse both expresses sounds and understands the signification of them, he is undoubtedly a living soul; and, being a living soul, composed by the omnipotent power of the Holy Spirit in concord with the design of God, therefore the said horse can never cease to exist in the exact form in which he was conceived and brought forth upon the stage of action, neither can he ever cease acting out the horse sense which the combination of simple elements in the spiritual pattern of the prototype of his species gives him. As he was conceived and born a horse, so will he everlastingly remain a horse, acting out the part of his species in the great drama of intelligent harmony.

Who has not seen that educated horse of the

circus, and having seen the appropriate manner in which he plays his part, can attribute it to any other cause than the intelligent action of an immortal soul? Most people have seen that troop of educated dogs, with a spotted clown, who plays his comic part with more pleasing variations than his biped rival, even to the funniest grimace and swaggering gait. Is it not wonderful to see how attentively each one of that quadruped troop, from his perch in the chair, watches for the point in the play at which his part comes in, descends and performs it, and returns to his seat, as proud and happy at the cheering of the audience as their rival bipeds of the ring, not to mention their depression of spirits at a silent reception of their acting and actual distress at a hiss. Those little quadrupeds are wonderfully amusing in the acting of their parts, which they perform with as much intelligent precision, amusement of the audience, and enjoyment to themselves as the bipeds of the ring do theirs, whom most people suppose to have souls.

As there is no wire-pulling behind the scenes in the performances of those little quadrupeds, but each acts his part with the same sense of power in himself to please the audience that his biped rival does, to what other cause can we attribute his performance than a knowledge of the effect which a certain act will produce upon the audience, and an artistic performance of the act to please, and draw therefrom demonstrations of approval of his capacity as an actor?

Thousands of instances are recorded in which dogs have saved the lives of their masters by a correct comprehension of the surrounding circumstances, at great peril, quick discernment of the only means of escape, and prompt rescue in the most fearless, determined manner, and by the only possible means. Let him who says it was not the act of an individual immortal soul, guided by the light of a correct process of reasoning to a sound judgment, executed by a fearless will, explain to us how else it was done.

To the catalogue of the record of cases of the prompt rescue of life from imminent peril by the sensible act of a quadruped we will add one more, that came under our own observation.

When about nine years of age, some five or six of our school-mates, ranging from six to ten years old, on a warm summer's day stripped and went into a mill-pond to bathe; as none of us could swim, we were wading and paddling about in shoal water, when one of the boys about eight years of age stumbled and plunged off into deep water and came up entirely out of reach of any assistance we could render him, at which our united force was expended in a cry of despair. A common uneducated cur dog, which had followed the then drowning boy from his father's house and taken up his watch on the bank, comprehended the perilous state of affairs at once and plunged in, and the instant his little master's head made its second appearance at the surface seized him by the hair, raised him up, and caught a firm hold of his arm near the shoulder, thus keeping his head above water, paddled him up to where we could reach him, and bore him out upon the bank, and by thorough rubbing and rolling and the fond caresses of the dog we soon restored him to consciousness. That dog was four years younger than his master, yet he saved his life by his own knowledge, sagacity, will, and intuitive force of character, without any dictation from us boys; for the first thought we had of him was his plunge into the deep water to the rescue of his master, four years older and larger of stature than himself, and was as happy at seeing him restored to consciousness as his master could have been at the rescue of his brother's life, while he would incessantly fondle him with his sensitive nose and lick the damp of death from his hands, face, eyes, mouth, and nostrils, and whine with the most anxious solicitude, and yelp and bark and even growl to rouse him from the stupor into which the process of drowning had prostrated him.

One shepherd dog is worth two men in herding, guarding, and protecting a flock of sheep. It is wonderful to see the skill with which they will urge forward a flock of sheep over pathless plains, head off the stragglers, bring them back into the fold, and keep the entire flock on a given course from camp to camp.

In Australia there is a species of wolf called dingoes, a sort of connecting link in the chain of animal life between the wolf and the dog, that the inhabitants of the island sometimes call wild dogs, which live in tribes, and have their hunting grounds as regularly laid off as the wild Indians of America had when the continent was discovered by man. If one of these tribes poaches on the hunting grounds of another it is sure to be chastised, and some heavy battles are fought in defence of the old landmarks; and when a herdsman ventured into the district of one of those tribes, with his flocks of fat mutton, the raids of the tribe of dingoes, on whose grounds he was feeding, were so systematically made, that in spite of all the shepherds could do, their losses amounted to hundreds of head per year, and in some neighborhoods numbered thousands. Under this high feeding the dingoes increased so rapidly, and the shepherds' losses in equal ratio, that the herdsmen were compelled to organize a standing army in defence of their flocks, and wage a war of extermination against the dingoes, which has been going on for a long time with varied success—the dingoes having destroyed about twenty sheep for each dingo captured by their enemies. Their tenacity of life is very great, and it is said that when captured they will feign death, and one is said to have maintained his feint till he was partially skinned, before showing any signs of life, when all at once he bounded off with a speed that gave the dogs a lively chase to recapture him.

Behold the lion, when he comes forth from his den to seize the prey which his own wants and those of his whelps demand, with flowing mane, steadfast purpose, and paralyzing gleam of eye. When the thunder of his war-cry peals over the plains, why does even the royal tiger crouch, trembling in his lair, till the actual locality of his terrible rival is known, and then flee in the opposite direction with all the power that is in him, to avoid an encounter with so terrible a foe?

If the voice of lightning is fuller in its volume as it peals over the plains, the vibrating death-knell of the lion is more appalling to both man and beast. If the burning ball of electricity is irresistible, the fatal grasp of the lion is no less fatal to animals; if its flash is more vivid, the angry glare of his eye is more terrible to encounter. The terror of all beasts, and undisputed monarch of the forest, he roams from jungle to jungle, and knows no fear.

But the skill of the hunter sends a bullet through the organs of thought, judgment, and will in that self-reliant head. One terrific bound— one desperate sweep of those huge paws in a vain effort to tear the very earth from its center, and down goes the carcass of that fearful monarch of the forest, stark, by the huge rock on which he has so often gamboled. A few spasmodic surges, convulsive tremors, and he stretches himself on the ground an immovable mass of terrestrial matter! Those gleaming orbs are glazed and sightless; those terrible limbs are stiffened with the chill of death: still, even that lifeless frame is an admirable statue of animal force and unquestioned courage, and his slayer approaches even his lifeless corpse with fear, and springs back at the slightest tremor of his departing life.

What made his voice more terrible than thunder—his spring more fatal than its bolt? and where has it gone? Let him who answers, "Instinct," bear in mind that he came upon the stage of active life through the same door of copulation that his wily conqueror did, worked at his own destiny with equal persistance, less fear, and made his exit through the same dark curtain which shuts the departed soul of man out from the view of terrestrial eyes:—in mein more majestic, anger more terrible, and in warfare more chivalrous than the man who slew him, and was only slain by the superior strategy of his wily foe.

Since the departure of the soul the intelligent motive-power which was driven out of that muscular structure of its own construction by the derangement of the machinery of the mind on which it operated and performed those appalling tenor strains in the great drama of life, that terrible structure of animal matter is as harmless as a marble statue, and is soon decomposed by the chemical elements which surround it.

For an hour after its departure the carcass remains warm and pliable: every limb is perfect; not a muscle of the body is injured; only the organ of will is unstrung and the spiritual operator departed. Such an operator! Is his knowledge obliterated? Has a leaden missile annihilated a sensible creature of Jehovah's composing, annulled a decree of the Almighty, and decomposed a celestial volition? Or has it only released an immortal soul from the prison-house of a terrestrial body, and given it a passport to the sublime joy of its eternal existence?

If a lion has no soul, what is it that looks out through those gleaming eyes and appalls every beholder? What force wields those terrible paws and sounds the thunder of his war-cry? Why do kings engrave him upon their banners of war as a badge of unconquerable majesty? If the eagle has no spiritual ambition, why do republics bear him upon their ensigns as a badge of soaring liberty, an emblem of liberty to the people?

Volition is as much the property of soul as color is of light. Consequently a design could no more be conceived and executed without a soul composed of a celestial body and spiritual operator, harmoniously working to one purpose, than a ray of light could be sent forth from the sun without a harmonious blending of equivalent parts from all the simple colors.

The voluntary movement of the body of an animal is as much the effect of a spiritual operator on the brain of the structure which was composed of celestial substances by an immutable law as vocal language is a harmonious combination of sounds uttered by the organs of speech.

Without a celestial machinery of the mind, and a spiritual operator, by which a will could be formed, no voluntary movement could be made, even by a monad, much less the terrific bound of a lion or soaring flight of an eagle, any more than a syllable could be pronounced, word uttered, or sentence addressed to the ear without the organs of speech, or instrumental music performed without an instrument.

Verily those elements of celestial substance and spiritual action out of which the soul is composed are all immortal, and being combined by an immutable law of affinity, fixed by an irrevocable decree of an Infinite Creator, in just that equivalent proportion which constitutes the specific structure of a lion, determines his power of action and mood of enjoyment. Therefore the patriarchal pair of lions so formed, together with each succeeding specimen propagated by them or their successors, commence an eternal round of lion-like volition at the moment of their conception, and from the day of their birth continue one eternal round of actions which fill the specific measure of their everlasting enjoyment; and as their species increase by copulation, and disappear 'from the earth by the dissolution of their terrestrial bodies, their souls join their predecessors in their celestial sphere, and form a social community whose fraternal sympathies harmonize in an immense group of social enjoyment, for the use and occupation of which God hath prepared them a habitation in the starry heavens, where there is plenty of room for all the increase of each pair of souls which he made in the beginning.

The structure of the lion, whether represented in statue by the sculptor in marble or bronze, engraved by the artist, his own carcass stiffened in death, restless movements in a cage, or seen in all the majesty of his movements on his native plains, is a pleasant study: one of nature's volumes, which strikes the senses so favorably that we love to read it. Such unwavering self-reliance in every feature; such a perfect model of strength, symmetry of form, and agility of movement; irresistible in will; and when we hear the appalling thunder of his voice, and see the terrific bound with which he secures his prey, we wonderingly exclaim, if thou hast no soul, whence come thy power and knowledge? If there is no spirit in animals, what is thy motive-power? What force directs the majesty of thy movements?

When God determined to make the king of

beasts—one whom all should stand in awe of, and none should willingly join in battle with, who should by unanimous consent wear the belt of superiority in animal strength, from whom even the royal tiger and elephant should flee in terror—he first conceived in his own mind the exact form of the head, size of brain, proportion of neck, shoulders, body, and limbs, length of mane, color of hair, size and color of eyes, structure of bones, twist of muscle, tension of nerves, exact proportion of every vein and artery, size of heart, liver, and lungs, shape and coil of intestines. Also the exact proportion of the simple elements of spirit, and equivalent arrangement and combination thereof, to compose the spiritual pattern of that majestic design, with a capacity to produce a routine of ideas and majestic movements concurrent with the majesty of the structure. Also the specific arrangement of such proportionate parts of the seven simple elements of the forces of growth as were equivalent to the production of the body in accordance with the design of the spiritual pattern which the Holy Spirit collected from the elements of nature, through the system of nervation, and combined into a concurrent mental operator and self-composer of its own substantial structure. Thus harmoniously corporated, they were infused into two ovoids, which had been maturely prepared in the composing womb, filled with the concurrent elements of celestial substance and terrestrial matter, to constitute the substantial form of its corporeal person in perfect concord with the spiritual pattern, where it commenced the work of life first by composing its own body, which was performed by the force of growth, operating from the heart, until it was worked up to a sufficient state of completion to admit of the operation of the spirit of intelligence upon the brain, whereby the nervous system could be tuned, a sensation produced, the muscles expanded and contracted, movement of those wonderfully-made limbs produced, and the body directed into the proper mood and intellectual sphere of a lion's enjoyment, at which time they were brought forth from the creative womb, male and female, and dropped in a jungle of the wilderness, in a climate suited to their nature, in a rock-bound cavern by the side of a spring of clear sparkling water, and plenty of rabbits and other small game in the vicinity to sustain them till they were sufficiently grown to go forth and fight the battle of life and fill even the tiger with fear.

That first pair of the family of lions, that prototype of the species, was just as perfect a model of the majestic structure of a lion as ever has been or ever will be produced by propagation among the family of lions. The spiritual pattern of that pair of lions, as drawn, modeled, and composed by an omniscient God, and combined into a living soul by the immutable force of the Holy Spirit to be the monarch of the forest and master of the quadruped group of the animal kingdom, was the perfect model of its type, the omniscient idea in its greatest perfection; and can never be improved upon or changed in a specific feature of its structure. A specimen may be dwarfed or even crippled into deformity by the disease of its mother in embryo, starvation in infancy, or casual injury: still every beholder knows it to be a dwarfed or deformed lion. It cannot assume the form of a tiger, wolf, or any other animal; and if its mother is in good health, and surroundings favorable, it will grow up to be a majestic type of the spiritual pattern. Thus we see a brief sketch of the mode of composing the spiritual pattern and living active souls of the first pair of each species in the animal kingdom, from the species of lions to mice, from man to a monad; and he who vainly imagines that he can make any improvement on the typical pair of any species of animal, in structural appearance, mental capacity, or mode of enjoyment, by any manner of treatment, or that he can compose a new species by amalgamation, will find all the labor bestowed to either of those ends as illy requited as the efforts of those alchemists were who vainly strove for some centuries to convert the baser metals into gold. We can take a family of any species, which has been reduced to an abortion of the vigorous spiritual pattern of its prototype by abuse and starvation in an uncongenial climate, and, by humane treatment and plenty of proper food, work it up to the standard of its prototype, but never surpass it. Is man wiser than God, that he should improve on his spiritual designs, add to the perfections of the drawing of his patterns, or change the specific structure of the souls which he has composed? Nay, verily, the work of man is to replenish, to care for all, bring all, even the lion, into a state of domestic order, and supply every species with its own specific food in such quantities as will keep them up to the perfect standard of the prototype; and the man who fails in this, and reduces any of his animals below the perfect standard of its prototype, by oppression of any kind, should be promptly punished by the government: starvation for starvation, exposure for exposure, and stripe for stripe. God has given man a stewardship over them, that both might be benefited thereby, and he will surely call every man to an account, and will punish each according to the abuse wherewith he hath oppressed the animals over which he hath control.

CHAPTER XVI.

SEVENTH GRADE.

FORCE OF GROWTH, AS ARRANGED FOR THE PRODUCTION OF BIPEDS.

The forces of animal growth, as composed by an equivalent arrangement of the seven simple elements for the propagation of this seventh group of the animal kingdom, do not vary so much in the size of their specific structures as they do in the quadruped group, none being as large as an elephant, strong as a lion, or small as a mouse. Still there are some very small specimens of the monkey, not nearly so large as a gray squirrel, and ranging from that species up to man, among which have been found very large specimens of giants, with a variety of intermediate species, differing greatly, of course, in natural propensities and personal appearances. Still they are all composed on the same structural idea of vertebrate bipeds. All of this group have but two feet, elongated pedals tipped with five toes, constructed in a manner to balance and support the body in an upright position, and most of them walk altogether upright, and all of them more or less so. They are also supplied with two hands, the extreme ends of which are composed of four fingers and a thumb, which are useful in many ways.

This group includes the whole biped species, from man to the lowest order of monkey, and is naturally divided into two grades, one of which embraces all the species of bipeds that feed at the table of nature, without replenishing or tillage, and live in a wild uncivilized condition; the other, only the family of man, whose intuitive propensities are to replenish, till, sow, reap, and gather into barns, build cities, carry on commerce, and enjoy life only in a mood of civilization.

This chapter will be devoted to the consideration of the former, and will complete the first volume of this work; and the second will be devoted entirely to the formation and structure of the soul of man, and his intuitive acts and propensities.

This first group includes every species of biped, from the wild Indian tribes to the lowest grade of monkey, all of which come under the cognomen of animals *feræ naturæ*. There is not one species in the group that will do a hand's turn in the way of tilling the earth, or use any effort in the way of replenishing, of their own free will; and there is but one species of the group that man has ever been able to domesticate or force within the pale of civilization that has ever been of sufficient utility to pay the expense of taming and educating, and that has brought with it and spread through the community a debauchment of morals and kindled a flame of social evil which in a few centuries has destroyed the nationality of the people who subjected them to servitude: so that up to this time this group of bipeds has been of far less utility in the great work of civilization than quadrupeds.

Their whole intuitive propensities run to fun, frolic, grimace, and trickery, and the whole group, as far as the work of civilization is concerned, are a useless set of vagabonds. Their entire mode of enjoyment lies in idle pleasure-seeking. They frolic and amuse themselves with monkey tricks and pastime enjoyments till hunger reminds them that they require food, when they go out and feed at the table of nature, if there is anything on it; if not, they die of starvation, and join their ancestors in their everlasting homes in the starry heavens.

Their sole object in life is to eat, amuse themselves, and increase the number of their species by propagation. Still they have no lust for copulation, except at that particular period in the course of nature when the female is in a condition to become pregnant and produce an increase in the family. No venereal diseases, brought on by hurtful indulgences contrary to nature, were ever bred among them. Their cup of enjoyment is caused to overflow by copulating at the proper period for propagation, and living in a simple come day go day God send plenty condition of generous plenty at some seasons of the year, and starvation at others, without ever dreaming that they could provide a regular supply by tillage and storing for the winter season: yet they are more honest in their dealings with each other and have less warfare between their tribes than men have hitherto had.

This group, like the quadrupeds, has no intuitive idea of a written language: still each species has a few words and many gestures, with which they communicate with each other; but their principal mode of exchanging ideas is by a spiritual reading of each other's thoughts, a specific communion of soul with soul, an affinity of mind with mind.

The individual members of each species live in much more intimate spiritual communion with each other than men do; they are almost free from that barrier of self-interest which severs the families of men from each other, and instigates all manner of deception and fraudulent tricks in their dealings between individuals.

The members of each tribe have no individual landmarks to create disturbance between neighbors; there are no line fences to be broken down, cultivated fields to be trespassed on, nor crops to be destroyed by their neighbors' animals; there is no aristocracy of pedigree or wealth, costly dwellings, or gorgeous dresses of new and rich patterns, to create envy, heart-burnings, and strife among them. They live harmoniously together, in strict subordination to the captain or chiefs of their tribes, who weigh out justice by the rule of equity, without fear, favor, or reward. They feed together on a common equality at the table of nature, and what is found thereon is divided equally among them, each taking for himself the required meal and no more.

The nesting places, bush tenements, mud huts, snow iglooks, or wigwams of each species are built precisely alike, and are of the precise pattern of the first habitation constructed by the prototype of the species; and all matrimonial alliances are sacredly respected among them, and the choosing of mates is the only source of individual contest. But when the marriage is consummated, it is sacredly respected by all parties.

The key-note of the intellectual harmony of this group in the tune of animal life is on B of the natural scale, appropriated to the soprano, and is flatted from the Indian tribes to the lowest grade of monkeys; and their part in the universal concert of the animal kingdom is principally performed by acting: there are very few musicians in it. Indeed, there is but one species of the group which has any talent for singing or whistling or instrumental music. With this exception, the whole group act their part in the great drama of life in the natural mood prescribed by the spiritual pattern which the design of the Almighty drew and fixed in composing the soul of the first pair of the species by the specific arrangement of the simple elements of finite spirit of intelligence and force of animal growth whereby they were produced and constituted living souls, in a similar manner to that by which the souls of the first pair of lions were produced.

This group naturally divides itself into the following grades: 1, baboons; 2, apes; 3, gorillas; 4, orang outang; 5, chimpanzees; 6, negroes; 7, Indians; in each of which is several different species, which never amalgamate, each increasing its own species in the spiritual pattern of their prototype, as constant as a tree bears seed after its kind that reproduces its specific pattern.

1. The baboon group consists of several varieties, all of which have long tails, which they use as a sort of grappling-anchor in their aërial flights from tree to tree; and it is said that one small-bodied species of them, which is pretty much all arms, legs, and tail, do their sleeping with their tails coiled around a limb of a tree and their bodies suspended beneath it, somewhat as fowls sleep on top of a limb with their toes clasped around it. One species of baboon are heavy-made and somewhat resemble a dog, and appear to be a sort of connecting link between the group of quadrupeds and bipeds. Although they are provided with a very respectable pair of hands, still they do most of their walking on all-fours, but sit upright and

use their hands very handily in feeding and doing many sharp little tricks.

It is said that the light-bodied long-tailed species of the baboon, or, as they are commonly termed, long-tailed monkeys, often migrate from one portion of their domain to another in tribes, and when they have a stream to cross construct a chain bridge, each link of which is composed of a baboon; in the construction of which they find two trees on opposite sides of the stream as near together as possible, and they go into the top of the one on their side so high, calculating the distance across, that a chain reaching to the ground will extend to a limb of the tree on the opposite side of the creek, when one baboon coils his tail around a limb, another takes a hitch around his body back of the arms, and another around him, and so on till they reach the ground, when the baboon at the bottom commences to swing the chain by striking his hands against the ground, giving an increased impetus to the lower end of the chain every time he comes into a position to do it, till he is able to grasp a limb of the tree on the opposite side of the stream at the same height of the other end of the chain, when the rest of the tribe, old and young, cross over on the chain bridge thus formed by the strongest of the tribe. As soon the last one is over, the hind baboon lets go his hold, and the chain swings across and uncoils from the bottom, and so the crossing of the stream is accomplished dry-shod. They are as afraid of water as a hen, and never venture to swim a stream, nor even to bathe or wash themselves. When infested with parasites, in place of washing to get rid of them, they hunt them very dexteriously with their slender fingers, and eat them, to get even for the bites they have suffered by them.

2. The most observable difference between the second grade, called apes, and the first grade, called baboons, is, they have no tails; and in their somersaults and leaps from limb to limb and tree to tree, they have to depend entirely on the hold they can catch and maintain with their hands and feet. Like the baboons, they live in the tops of trees and sleep among the branches. They suckle their young at the breast, and tend them with great care, tossing, handling, and caressing them in a manner wonderfully human. Their most prominent trait of character is a propensity to imitate whatever they see done. Thus we are told that a traveler in the country inhabited by them, sitting under a tree and observing their movements, finally laid down and fell asleep. On awaking, he found that his hat had been appropriated by one of the apes, who had perched himself in the top of a very high tree, with the hat covering his hairy head. After waiting for some time, in hopes that he would tire of it and throw it down, to no purpose, he finally fixed up

his handkerchief in the form of a hat, walked about till he attracted the ape's attention, and then took off his resemblance of a hat, gave it a flourish, and threw it down, at which the ape performed the same trick, and he recovered his hat.

Their propensities all run to merry-making and mischief, so that no person has ever been able to educate them to any useful purpose that would pay the expense of their keeping, which is perhaps to some extent owing to a lack of patience, tact, and perseverance in the process of education. We have no doubt some expert animal teacher will some time educate a class of them to perform some useful service, and raising the young under the same discipline will eventually submit to domestic order, and use their extraordinary skill, strength, and imitative genius in a manner that will save them from annihilation when the increase of population has brought the earth under a state of close cultivation.

3. The third grade of intelligence in this group, called gorillas, are the largest and most ferocious of these seven grades of bipeds. They are pugnacious of temper, and are of about equal strength with a bear, which they somewhat resemble when a bear rises on its hind legs; still they are decidedly biped, have a very respectable pair of hands, and walk upright. On account of their ferocity, both hunters and travelers avoid them;

consequently there is very little known of their moods of enjoyment or capacity for being taught to pursue any useful labor. It may yet be discovered that they are well qualified to render some useful service in tilling the earth under the kind direction of man, that will pay well for their own living even after the earth is all brought under a state of cultivation. As the country in which this lower grade of bipeds lives is uninhabited by man, and the only knowledge we have of their intuitive moods of enjoyment and intellectual propensities has been acquired through the representations of hunters, whose study is to destroy the ignorant defenceless animals, and not to make their acquaintance on terms of amity or study their natural propensities in their best moods, our knowledge of them has been so limited, that it was for a long time supposed that the gorilla was an adult chimpanzee; but zoölogists now agree that they are a different species, and have assigned to each a specific name.

4. The next grade is the orang outang. They are said to build their habitations in the tops of trees by weaving the branches together. They are neither so large nor ferocious as the gorilla, and are somewhat better understood; but as they have been chased, hunted down, and killed by hunters, till they have been educated to look upon man as their natural enemy, they are always in an angry or frightened condition whenever they

meet, and when captured and taken out of their native country, far away from kith and kin, partly through bad treatment, and partly through home sickness, and a longing desire for their own genial climate, liberty, and kindred associations, they become moody, sickly, and are short lived, so that we have really no correct idea of their intuitive propensities and home moods of enjoyments, domestic pastimes, mode of rearing the young, or providing for their family.

5. The chimpanzee: Concerning this grade of the group of bipeds, we will give an extract from the Natural History of the Rev. J. G. Wood, (vol. 1, p. 20:)

"Closely connected with the gorilla is the large black ape, which is now well known by the name of chimpanzee.

"This creature is found in the same portion of Western Africa as the gorilla, being very common near the Gaboon: it ranges over a considerable space of country, inhabiting a belt of land some ten or more degrees north and south of the torrid zone. For some little time it was supposed that the gorilla was an adult chimpanzee, but zoölogists now agree in separating it from that animal, and giving it a specific name of its own. The title negro or black sufficiently indicates the color. * * *

"In its native country the chimpanzee lives in a partially social state, and at night the united cries of the community fill the air. If we may credit the reports of the natives of Western Africa, the chimpanzees weave huts for themselves, and take up their residence in these dwellings. Now, it is a well-known fact that the orang outang, which comes next in our list, can rapidly form a kind of platform of interwoven branches, and so it is not beyond the bounds of credulity that the chimpanzee may perform a work of similar character, only the chief difference between the customs of the two animals seems to be, that the one lives upon the structure or roof, if it may so be called, and the other beneath it. Some travelers say, that although the huts are actually inhabited, yet that only the females and young are allowed to take possession of their interior, and that the male takes up his position on the roof. * * *

"It is a remarkable fact that the chimpanzees are groundlings, and are not accustomed to habitual residence among the branches of the trees. Although these apes do not avail themselves of the protection which would be afforded by a loftier habitation, yet they are individually so strong and collectively so formidable, that they dwell in security, unharmed even by the lion, leopard, or other members of the cat tribe, which are so dreaded by the monkey tribes generally. Even the elephant yields to these active and ferocious animals, and leaves them undisturbed.

"Yet a chimpanzee would not dare to meet a panther in single combat, and depend for safety upon the assistance that would be afforded by its companions. This was shown by a curious and rather absurd incident on board a ship, where a young and docile chimpanzee suddenly came in sight of a caged panther which had taken passage in the same vessel.

"The unexpected sight of the panther entirely overcame his fears, and with a fearful yell he dashed along the deck, knocking over sundry of the crew in his passage. He then darted into the folds of a sail which was lying on deck, covered himself up with the sail cloth, and was in such an agony of terror that he could not be induced to come out of his retreat for a long time.

"His fright was not groundless, for the panther was as much excited as the ape, only with eager desire, and not with fear. It paced its cage for hours afterwards, and continued to watch, much as a cat may be seen to watch the crevice through which a mouse has made good its escape.

"The food of these creatures appears to be almost entirely of a vegetable nature; and they are very unprofitable neighbors to any one who has the misfortune to raise a crop of rice or to plant bananas, plantains, or pawpaws within an easy journey of a chimpanzee settlement.

"Many specimens have been brought to Europe, and some to England, but this insular climate seems to have a more deleterious effect on the constitutions of these apes than any of the other quadrumane. In this country our most insidious and most irresistible maladies fasten upon the apes with relentless hand. The lungs of these creatures are accustomed to the burning sun, which heats and rarefies the air of the tropical climates, and they are peculiarly sensitive to cold and damp. Few members of this family live to any length of time after they have once crossed the channel; for after a while they are seized with a short hacking cough, the sure sign that consumption has begun that work which it is certain to accomplish. * * *

"A monkey, when afflicted with this disease, is truly a pitiable sight. The poor animal sits in such a woful attitude, coughing at intervals, and putting its hands to its sides, terribly human, and looks so mournfully and reproachfully out of its dark brown eyes, just as if it were rebuking the spectator for his part in bringing it from its native land, where it was happy among its friends, to die a solitary death of cold and consumption behind the bars of its prison.

"In the Jardin des Plantes in Paris there was a remarkably fine specimen of the chimpanzee, black, sleek, and glossy. He was facile princeps in the establishment, and none dared to dispute his authority. He was active enough, and displayed very great strength and some agility as he swung from side to side of the cage by means of the ropes which are suspended, but he preserved a dignified air, as became the sole ruler.

"There was a kind of aristocratic calmness about the animal, and he would at inverals scan the large assembly that generally surrounded the monkey-house. His survey completed, he would eat a nut or piece of biscuit, and leisurely recommence his gambols. * * *

"It is said that they have a sufficient amount of knowledge to be aware that the strength of a man lies in his weapons, and not in his muscles only; and that if a hunter should bring on himself the vengeance of the troop, by wounding or killing one of their number, he can escape certain death by flinging down his gun. The enraged apes gather around the object that dealt the fatal stroke, and tear it to pieces with every mark of fury. While they are occupied with wreaking their vengeance on the senseless object, the owner of the fatal weapon escapes unnoticed.

"Though the language of apes be not articulate, according to our ideas, yet in their wild state the chimpanzees can talk well enough for their own purposes. One proof of this is the acknowledged fact that they can confer with one another sufficiently to act in unison at the same time and place, and with a given design.

"Strong and daring as they are, they do not appear to seek a contest with human beings, but do their best to keep quietly out of the way. Like most animals that herd together, even in limited numbers, the chimpanzees have ever a watchful sentinel posted on the lookout, whose duty it is to guard against the insidious approach of foes, and to give warning if he sees, hears, or smells anything of a suspicious character.

"Should the sentinel ape perceive a sign of danger he sets up a loud cry, which has been likened to the anguished scream of a man in sore distress. The other apes know well enough the meaning of that cry, and signify their comprehension by answering cries. If the danger continues to threaten them, the ape conversation becomes loud, shrill, and harsh, and the air is filled with the various notes of the simian language, perfectly understood by themselves, although to human ears it consists of nothing but discordant sounds, yells, and barks."

We could not refrain from giving the above extract from that natural history; it is so appropriate, and sketches so graphically some of the moods and intuitive propensities of these lower grades of bipeds. What this entertaining historian says about their specific language and understanding of it is true of every other species of animal living. As they differ in specific appearance, so they differ in language, tribal organization, intuitive propensities, mood of enjoyment, and manner of working out their destiny of this life, for the reason that there is a corresponding difference in the specific arrangement of the seven simple elements of force which constructed their bodies, and necessarily a corresponding difference in the arrangement and combination of the seven simple elements of finite spirit in composing the spiritual pattern of the souls of the first pair of the species, whereby the whole race is molded and quickened into a specific family, every one of which will, from the day of its birth, help its family to act their specific part in the great concert of universal intelligence through the everlasting rounds of eternity.

Why all Christendom should have consigned the whole animal kingdom below the grade of a caudal negro to annihilation, and every creature above it to immortality, passes our understanding. Christ taught no such doctrine, and Peter testifies that he saw a certain vessel, as it were a great sheet knit at the four corners, let down from an opening in heaven, filled with the souls of all manner of beasts. Why should God exert his omnipotent force to compose the spiritual pattern of any species of animal with a capacity to increase the number of its family by propagation to an indefinite number, only to put him to the trouble of annihilating them again when they had only commenced acting their part in the everlasting drama of finite intelligence? We should always remember that when any one thing which God has composed ceases to exist in the specific form he gave it, that moment his creative power ceases to be infinite, for his infinitude can only be sustained by the infinite duration and immutable structural form and specific movement and action of his works.

And we should not forget that every new sensation that passes through the nervous system of any living animal, to or from the mind of which it is conscious, is a new note sung in the universal anthem of intellectual praise and thanksgiving to God for having endowed them with mobility, and a consciousness of its own feelings, and knowledge of surrounding objects. However inferior any creature may look to us, we should remember that God composed the spiritual pattern in the soul of its prototype by an equivalent arrangement of simple elements from the same department of natural element from which the soul of our prototype was composed; and that the same kind of atoms of spirit, mites of force, and molecules of matter, of which we are composed by copulative propagation, being differently arranged and combined in the soul of their prototype, now compose them by the same process of copulative propagation: that their form, whatever it may be, is as comely to them as ours is to us; and that their ideas, such as they are, fill their measure of joy just as full as our ideas do our measure of enjoyment, and, being designed and composed by the Almighty, fill their sphere in the halo of his glory as well as we do. Their structure is his design, their intelligence his endowment and perfectly satisfactory to them, and help to complete the glory of God's infinite designs; and their everlasting life helps to perfect the infinitude of his work; and their specific action helps to fill the anthem of praise and thanksgiving to God for his infinite goodness and mercy.

Consequently, should a single specimen of the most inferior insect, after having been born and commenced a sensational life of intelligent action, ever change that specific routine of action, the immutability of God's designs would be broken; and if it should ever cease to exist, the infinitude of his works would be impeached; aye, and convicted, too, of being finite in that instance. Hence Christ said, he suffereth not a sparrow to fall to the ground without his notice; and as the soul slips out of that little gasping body, he directs its flight to join the beatified souls of its ancestors in the celestial sphere prepared for them before the first pair were composed.

What appears to our terrestrial senses to be a construction and destruction of vegetable and animal structures, are only the mold of sand in which the celestial jewel is cast, the solidified structure of terrestrial gases whereby the immortal soul is composed, which is just as necessary an element for composing the soul by copulation as the mold is in casting the jewel, and the destruction of the mold is necessary before the jewel can be of any essential utility: so also the soul cannot attain to its full capacity of glorifying God till it has escaped from the dross of terrestrial gases in which it was composed.

The simple elements of terrestrial matter alone are subject to chemical synthesis and analysis, which are only used in molding celestial things and composing immortal souls, which, when completed, are decomposed to their elementary condition to be used again. But all celestial substances, forces of growth, and intellectual sprit, once composed into a soul, it is everlastingly immortal and specifically immutable as God himself from the day that it becomes a living soul.

6. The next grade in this group of bipeds is the races of negroes, of which there are several species. The first and most distinctly marked and next link in the upper grade above the chimpanzee is the caudal negro, found in the southern portion of Africa; they are more docile than the chimpanzee, and some of them have been captured when young, and educated to render a sufficient amount of service to pay the expense of their keeping. They are both carniverous and gramniverous; but their carniverous propensities are so strong that, unless they are fed a certain proportion of meat, they have been known to devour even the children of their masters.

In structure they are very little superior to the chimpanzee, while their skin resembles the negro proper, being void of hair and equally black, and their heads are covered with wool of a little coarser quality than the negro proper.

They are groundlings, and their huts are constructed partly of an excavation in the ground, covered with an arch, composed of balls of mud, laid up in a plastic condition, which, when dry, makes an arch of considerable strength: the place of egress and ingress is a hole at the bottom and partly under ground, just large enough to admit the body, somewhat resembling the hole of a badger leading to its burrow, in which they burrow in the winter, but in the summer they live mostly in the jungles, sleeping under trees.

They are natives of a temperate zone, and use their huts to protect them from the cold storms of winter, and the forest shades to protect them against the extreme heat of summer. Their hut is the only structure which their hands have ever made; they wear no clothing, not even a piece of stuff about their loins; and as far as manufacturing is concerned they are no way superior to the chimpanzee. He constructs his hut of brush, the caudal negro of balls of mud; and both feed at the table of nature on what Providence sends them, and that is the end of their work. The former is covered with hair, while the latter has not a hair on his body. The former has no tail at all, while the latter sports an elongation of the back-bone somewhat resembling the tail of a sheep. The former feeds upon vegetables, while the latter gorges himself from the carcass of an ape, negro proper, or even a man, with great satisfaction.

He is susceptible of being taught to perform some useful services under the supervision of man, as an English doctor at the Cape of Good Hope has proved, by capturing a young one and carefully educating it to render certain services; but his own desire for improvement never rises above his mud hut, wild life, and such food as Providence sends him.

Besides this caudal variety there are several other species, most of which inhabit the torrid zone; the next above the caudal negro, whose range extends partially into the temperate zone, are hardly as large in stature as the tailed negro, and are quite as awkwardly made, and there is about the same difference between them, that there is between a long-tailed and no-tailed monkey, the principal difference being, one has a tail and the other has none, and their mode of working out their own salvation in this life is very similar, but being south of the slave markets, very few of them have ever been educated to perform any useful service: they are very nearly as stupid as the caudal negro.

All of the higher grades of negroes inhabit the torrid zone, and however hot the sun may shine, it never seems to be quite hot enough for them. Their greatest delight is to lie in the most exposed

places and bask in its scorching beams. They stand
he heat better even than the chimpanzee, and seem
to have been designed to inhabit the hottest and
most sickly miasmal districts of the torrid zone in
Africa.

The principal difference between the structure
of their mud huts and the huts of the caudal negro
is, they seek a timberless plain, void of shade, on
some rich alluvial river bottom, and construct
them on top of the ground with balls of black
adobe, in the form of a conoid, somewhat resembling the inverted nest of a hornet. The opening
for egress and ingress is at the bottom, a hole just
large enough to admit the body. These black
adobe structures, exposed to the vertical rays of a
tropical sun, become so hot, that an hour's residence in them would kill an ape, in which, like
salamanders, they live, copulate, and raise their
young. They live in the same state of nudity as
the caudal negro, and their mud hut is the end
of their building capacity; and having no desire
for clothing of any kind, their intuitive manufacturing propensities are in no respect above the
caudal negro.

They are both carniverous and gramniverous,
feeding partly upon such grasses and vegetables
as the earth naturally produces, and partly upon
monkeys and such other animals as they are able
to capture; and missionaries and travelers among
them inform us that one species of negro feeds

greedily upon the carcasses of other species of negroes, tearing the flesh to pieces with their teeth and fingers, having no more idea of the manufacture of cutlery or other household furniture than the apes.

When well fed they are docile and easily captured, and are capable of being educated to render useful service in many ways. In the use of the spade, mattock, and hoe they soon learn to perform good service, under the constant care and direction of a driver. This was discovered at an early day by the Egyptians, Moors, and other African nations who lived north of them; and as they lived in tribes like the chimpanzee, and built their mud huts in clusters in sufficient numbers to be a mutual protection to each other against the depredations of other species of negroes, lions, tigers, and other carniverous beasts which inhabit that country, they were an easy prey to the neighboring nations of men, who by kind treatment and friendly approaches succeeded in maintaining a sort of friendly intercourse by making them a few presents of beads, or some worthless ornaments for their naked black persons. Thus they created a desire among them for brass earrings and rings in the nose, and brass bands about the wrists, and beads about the neck, to such an extent that the parents would sell their children into slavery for a brass ring or string of beads that was not worth six cents.

Like all other animals, their paternal care over their helpless young is very strong, but weakens rapidly as they approach maturity and become capable of caring for themselves.

The worst feature of this traffic was, those unscrupulous traders become very intimate with the wives of the chiefs of the tribes, and perpetrated amalgamation with them, whereby the spirit of a man was imprisoned in the body of a negro, which it abhorred; and as the hybrid grew up to maturity, endowed with all the bad qualities of both races, it became an incarnate devil, possessed with the activity and perseverance of the soul of man, clothed in a body which it loathed and hated, and doomed to dwell among a race of animals which it despised, and for whom it had neither fellowship nor fraternal sympathy; and its own members being at enmity with each other, there was not a vestige of human sympathy left in its composition, and its sole desire was to wreak its vengeance on all the race among which it was compelled to dwell.

These hybrids being vastly superior to the thorough-breds in designing capacity and strength of will to execute any purpose, and being the sons of the wives of the chiefs of the tribes, naturally became the ruling element, and were thus enabled to practice upon them the most diabolical barbarities, and were soon able to organize a small army of hybrids, who were always ready to go out to a neighboring tribe and capture as many as any slave dealer was willing to buy for trinkets and things of no value.

Notwithstanding the spirit of a man was in them, they were clothed with a carcass which loathed the food of civilization, and craved only that which Providence furnished at the table of nature, so that there was no occasion for using their forces in any civilized pursuit; still they were bound to work off their excessive forces at something, which was usually done in warfare with some neighboring species of animal. A very few days' rest made them feel so happy, that they were again ready to start out on a raid, and kill somebody and capture some prisoners from neighboring tribes, who, in case there were no slave buyers in market, became an expensive trouble on their hands, which to get rid of they would kill, cut up, roast, and make a barbecue feast for the tribe.

Thus by the sin of amalgamation those once peaceful happy tribes were converted into warring, devouring demons, and subjected to the most terrible oppression, liable at any time to be captured and sold into servitude, or slaughtered and gobbled up by their cannibal neighbors. The sin of amalgamation did not stop here with the native tribes, but followed the captives into servitude, where nearly all the children of the wenches were begotten by their masters' overseers or interloping vagabonds, and thus the country in which they were subjected to servitude became populated to a considerable extent with hybrids, who retained all the bad qualities of both species, composed of the active energetic soul of man imprisoned in a gross carcass, which required gross food, and was the most comfortable in a state of nudity, and enjoyed itself only in a gross state of moral turpitude.

The sexual effect of the cross was a breaking up of the natural courses of nature, so that the females were always in heat and excited to an inordinate desire for copulation, that stripped them of all feminine modesty and virtuous sense of the proper use of copulation, and filled their whole nervous system with a burning desire for the sexual embrace at all times and under all circumstances, under the influence of which they were constantly exposing themselves in tempting attitudes to the males of both whites and blacks, and using every seductive art that they were master of to induce copulating indulgence, without any regard to matrimonial alliances, which excited all the evil passions of men to a state of reckless riot that broke down all virtuous restraint, and led to such an unnatural use of copulation, that loathsome venereal diseases were engendered, the blood loaded with virus matter, prostrating every vital energy, eradicating all sense of virtuous propriety and manly aspiration for human progress, and dragging the whole community into a state of barbarism far beneath the aboriginal thorough-bred negro tribes; for what little energy they had left from forces expended in sexual abuse was expended in warring with, killing, and devouring each other, constantly raiding upon neighboring nations to obtain supplies which they were too indolent to produce for themselves. Consequently some neighboring nation, which had a little regard for virtue, care for its own personal health, and desire for progress in civilization, was compelled to subjugate them and force them to a passive observance of the laws of civilization.

Thus we see that Arabians, Egyptians, and Moors, and all those African nations which first introduced negro slavery among them, with all of its corrupting vices, by hybridizing and whoredom, have for thousands of years been only groveling abortions of manhood—mere fossils of the national dignity to which they attained in a century under the reign of a virtuous manhood, unadulterated by amalgamation with inferior races, nor prostrated by venereal disease, brought on by an unnatural use of copulation. Most of those countries which were once in a high state of cultivation and densely populated by industrious, thriving, happy people, are now a wilderness, inhabited only by hybrid bandits, presenting every shade of miscegenary abortion, from negro to man, who live at the table of nature on what

Providence sends them, except what they can get by plunder and robbery, and have destroyed every vestige of civilization within reach of them, through which no man can travel without military escort: and Mexico and Central and South America are already in a similar condition by the same cause; and a ruling dynasty of hybrids will produce the same result in a very few years in any country in the world.

The negro races, as God composed them, are an innocent, harmless, happy people, who when left to work out their own destiny, live just as God designed they should live, in the hottest, unhealthiest part of the world, in a perfect state of nudity, in as wild and uncivilized a condition as the monkey tribes, where Providence feeds them with food exactly adapted to their desires and most conducive to their health and welfare, that they may increase as rapidly as possible for the population of the sphere which God prepared for them in the starry heavens ere he composed the spiritual pattern of the souls of the first pair of each species of negroes, the males of which have never marred that pattern, nor disgraced their species by amalgamating with any other race, whether of other species of negroes, apes, Indians, or men; and the souls of each species of negro, whether of the tailed or no-tailed species, has invariably, from the time of its creation, worked out its own salvation in the precise manner that God designed it should when he composed it, and has been as constant to the spiritual pattern of its prototype as a chimpanzee: and in its simple wild life, if it does no more good than an ape, it certainly does no more harm, and the sin of enslaving them and forcing the females to submit to amalgamation has been visited with certain death, national and moral, upon every nation that has perpetrated it.

Mark, now, this whole brood of hybrids in the world is the result of a cross between man, squaws, and negro wenches. The soul of a negro will not disgrace itself by copulating with a woman: the entire responsibility of this sin rests on the soul of man, and terribly has man suffered for it If there is any creature on earth which deserves all the pity which can possibly be given to it, it is a hybrid, a mulatto, composed of the soul of a man, imprisoned in the carcass of a negro, which it hates and loathes, by the accursed lust of its father. But pity does him no good; for nothing can relieve him from the body of that moral death but death itself. Hence the Divine law commands that he who perpetrates that sin shall be stoned to death, and that the mother and the offspring shall also be put to death, that the innocent hybrid may be saved from a life of shame, a living death. We believe that if our father had begotten us of a negro wench, if she would tell us who he was we would follow him to the very gates of hell to see that sentence of stoning to death executed upon him; and if the government would not do it, we would execute it with our own hands, and God would justify the act.

When God determined to compose the spiritual pattern of the soul of the primitive pair of negroes, he conceived in his own mind just the change in the arrangement of the simple elements of finite spirit and force of growth that would remove the caudal negro one grade above the chimpanzee; and being thus constituted, the soul of the caudal negro has been just as constant to its spiritual pattern in propagating its race to the present time as the lion has to the spiritual pattern of its prototype. And thus was the next grade of negro above the caudal constituted, and thus the next above that, and so on till the highest grade of the negro race was reached, each of which has been just as constant to the spiritual pattern of their prototype as the various races of baboons, apes, and chimpanzees have to theirs, never having amalgamated with each other in any instance. One species of negro, when hard pressed with famine, will kill and feed upon another species of negro, but never amalgamate. Trees in the forest were never more constant to their species than the negro races are in repeating their spiritual pattern by propagation; and from the highest to the lowest of them, they live in the same wild state of nudity that the apes do, nor have they any more manufacturing talent. An idle, merry life is their constitutional status, which they could not depart from without breaking the designs of the Almighty.

Some of these races have quite a talent for music, and both sing and whistle a native carol of song in a very entertaining manner, and are as earnest and happy in it as the birds of the air; and when left to themselves, uninfluenced by the designs of man and uncorrupted by his lusts, they work out their destiny of life with more constancy to the designs of their Creator than man has ever done up to this time.

7. The specific range in tone between the performers of this scale of the soprano part of the tune of universal intelligence rises by the same simple even tone from the highest grade of the negro races to the lowest grade of the Indian races that it does from the highest grade of the ape species to the lowest grade of the negro races.

The Esquimaux are undoubtedly the lowest grade of the Indian races, who in point of intelligence and intuitive work of life range just one simple tone above that of the highest constructive intelligence of the highest grade of negroes which the rigor of the climate in which they live compels them to perform; for while there is just one simple tone's variation between the intelligence of the Congo negro and Esquimau Indian, their habitations are antipodal, *i. e.*, the negro lives in the torrid zone, in a mud hut, built in a miasmal bottom, exposed to the most intense heat of the vertical rays of a tropical sun, void of any covering for his body but that which the Creator gave him, and stands, aye, and enjoys, the most intense heat equal to salamanders. Whereas the Esquimau Indian lives in the frigid zone, environed by perpetual snows, builds his igloo of snow, sleeps on a snow bed, and lives entirely on such meat as the sea furnishes for his sustenance. And as the Creator did not provide him with a coat of warm fur to shield him from the rigor of his snow bed, frigid climate, and snow hut, as he did the polar bear and seal, he endowed him with a capacity to provide one for himself, which he does by stripping off the skin of a bear whole, except an enlargement of the slit at the mouth to an extent that will admit of its being drawn back over the body entirely whole, except the enlargement at the mouth, and tan it into a pliable condition by moistening it with the brains of the bear and then rubbing and beating it dry, moistening again with brains and then beating it dry again, and so on till the horny stiffness is taken out of it, when he slips his biped legs into the skin of the hind legs of the quadruped, and his arms into the skin of his fore legs; and their structures are so near alike that a tailor could not give him a better fit. They also prepare fur blankets for their snow beds by tanning seal skins in the same manner, and then sowing them together with a bone needle and strips of seal skin or ligaments from the muscles of the bear. In this one thing they excel the highest grade of the negro race in manufacturing intelligence by intuition.

In capacity to endure heat the negro races partake of the nature of salamanders, while the Esquimaux seem to be equally allied to the polar bear and ice-bound seal in constitutional ability to endure cold. While one hundred degrees Fahrenheit is the lowest temperature congenial to the negro races, zero is the highest that is congenial to the Esquimaux. One day and night in a negro's mud hut, under the burning sun of the torrid zone, would not only do an Esquimau to death, but would try out the seal oil and whale blubber from his carcass, and leave only a skeleton covered with a mummy-like skin. So, also, one day and night's lodging in a snow igloo, on the snow bed of an Esquimau in the frigid zone, would freeze a negro as stiff as a statue of black marble. Thus God graciously made just such an arrangement of the simple elements of finite spirit, in composing the spiritual pattern of each of those species, as qualified them to enjoy life in the climate in which they were destined to live, and fatten on the food contained in the natural products of the country, and endowed each with just the degree of intelligence that would enable him to laugh and grow fat in the environments allotted to him, void of regret for the past, reckless of the future, and enjoying the present with a merry zest to which man seldom attains.

We will now give a few extracts from C. F. Hall's Life among the Esquimaux, which will illustrate the Esquimaux mood of enjoying life:

"The next thing we did was to build an igloo, where at all events something like a shelter could be obtained and warmth by clustering together. * * * Koker Johin, the master mason, aided by Sterry, built the igloo out of a snow bank. * * * The igloo completed, on lying down we found it was too limited, and that we should be inconveniently, and perhaps injuriously, cramped; therefore a remedy must be found, and this was by cutting pigeon holes in the snow bank for our feet. This answered, and soon we were fast asleep, though upon a bed of snow, and at my back a snow bank. * * * (p. 264.)

"The island we were now going to was the one where Annawa and his family went to at the time we escorted them part of the way the previous fall, (p. 128,) and we now intended to rest there for the night; but it was quite 9 p. m. before we arrived, and then some of the family were in bed. This, however, did not prevent our having a prompt and most friendly reception. The aged Annawa and all those with him quickly gave us food, and rejoiced to see me; and, though there was no spare bed, yet I was cordially invited to share theirs. Soon after, tired and sore with my long walk of more than twenty miles over ice, mountain, and ice again, I retired to rest as best I could.

"That night my sleep was a sound one, though I was tightly squeezed, the sleepers being numerous, and all in the same bed. There were nine of us besides *the infant* at the breast—a boy three and a half feet in height of portly dimensions. The order of sleeping was as follows: Key-e-zhune, the wife of Annawa, lay in her place by the ik-kumer, or fire light, with infant, Kok-uk-jun, between herself and her husband; then, next to him was the child Oo-suk-jee. I lay alongside of the child Koojessee. Next to me there came Esh-ce-loo, with his wife, Oony-a, all of us facing upward. Then, with feet at our feet, were a young man, Innuit, and the little girl, Kimmiloo, who lives with Annawa.

"The space into which the ten were compacted and interwoven was less than as many feet. Of course I had to sleep in my day dress, as no spare bed is kept in reserve for company, nor had they more covering than they needed for the family; but I got along through the night after a fashion.

"It was, however, not very pleasant. Whenever I attempted to turn to relieve my aching bones, a little boy by my side roared like a young lion, waking all the sleepers, and thus creating a confusion that would have deprived me of further slumber but for my great fatigue. However the night passed, and early in the morning I slipped out, as a snake from deciduous epidermis, and prepared for a walk. * * * (p. 258.)

"Our breakfast and dinner were excellent. For the former, raw frozen walrus, of which I had a piece for my share of about five pounds; and at the latter seal. The portion of this allotted to me and Sterry was the head. We complied with the Innuit custom. Sterry took a mouthful, then passed it to me, and when I had done the same, it was returned to him, and so on. Of course fingers were all in all. No knives are found among the Innuits; fingers and teeth are more than their equivalent.

"When the meat, skin, and hair were all despatched—even the eyes, except the balls, which were given to the youngest child of Sampson— we tapped the brain. I was surprised at the amount of a seal's brain, and equally so at the deliciousness of them. The skull was almost as thin as paper.

"Later in the day I attended another feast in the igloo of Kookin, who had invited his old mother, Shel-lu-ar-ping, and two other venerable dames; and I must say, that if my friends at home could then have seen how like an Innuit I ate, they would have blushed for me.

"First came a portion of seal's liver, raw and warm from its late existence in life. This, with a slice of ook-sook, (blubber,) was handed to each, and I made way with mine as quickly as any of the old adepts. Then came ribs, enclosed in tender meat, dripping with blood. How ambrosial to my appetite! Lastly came—what? Entrails, which the old lady drew through her fingers yards in length. This was served to every one but me in pieces of two to three feet long. I saw at once that it was supposed I would not like to eat this delicacy; but having partaken of it before, I signified my wish to do so now: for, be it remembered, there is not any part of a seal but is good. I drew the ribbon-like food through my teeth, Innuit fashion, finished it, and then asked for more. This immensely pleased the old dames. They were in ecstacies. It seemed as if they thought me the best in the group. They laughed; they bestowed upon me all the most pleasing epithets their language would admit. I was one of the honored few.

"Soon as the round of feasting was ended, one of the old lady Innuits drew my attention to her afflictions. She had a dreadful pain in her side and back, and had been badly troubled for weeks. Before I had time for thought she drew off her long-tailed coat over her head, and sat there before me nude as nature made her. The laughing face and joyous, ringing voice of the old lady were now exchanged for expressions indicative of suffering and the need of sympathy. The whole party present were now absorbed in the subject before me. I put on as long and dignified a face as I could in this trying scene, and as much was evidently expected from me, I was determined no disappointment should follow. Therefore I proceeded to manipulate the parts affected, or rather plowed my fingers in the rich loam (real estate) that covered the ailing places. The result was, that I gave notice that she should live on eating as much fish, seal, and walrus as she wanted, drinking water several times a day, and applying the same amount at the end of every ten days as she had drunk in that time to the outside of her body by the process of scrubbing, which I then and there practically explained to her and the others. * * * (p. 266.)

"When a white man for the first time enters one of their tujoes or igloos, he is nauseated with everything he sees or smells—even disgusted with the innocent natives, who extend to him the best hospitality their means afford. Take for instance the igloo in which I had an excellent dinner on the day last mentioned. Any one from the States, if entering this igloo with me, would see a company of what he would call a dirty set of human beings, mixed up among masses of nasty uneatable flesh, skins, blood, and bones, scattered all about the igloo. He would see hanging over a long flame ('stone pot, filled with blubber burning from a moss wick') the oo-koo-sin, (stone kettle,) black with soot and oil, of great age, and filled to its utmost capacity with black meat swimming in a dark smoky fluid, as if made by boiling down the dirty scrapings of a butcher's stall. He would see men, women, and children—my humble self included—devouring the contents of that kettle, and he would pity the human beings who could be reduced to such necessity as to eat the horrid stuff. The dishes out of which the soup is taken would turn his stomach, especially when he would see dogs wash them out with their pliant tongues previous to our using them." (p. 522.)

Thus we have their progress in architecture and the culinary art after some six thousand years, in which their species have lived, experimented, and passed away, and something over three hundred years they have had the example and instruction of missionaries, traders, and sailors, who have been much among them.

"I could here mention one or two facts, but it will be unnecessary more than to say that mothers here at home will comprehend all my meaning when I tell them that Innuit infants are carried naked in the mother's hood." (p. 180.)

"It has been said by a well-known witty writer, now deceased, when referring to the Esquimaux, in an Arctic book he was reviewing, that they are singular composite beings—a link between Saxon and seal—hybrids—putting the seal's body into their own, and thus encasing their skins in seals; thus walking to and fro a compound formation. A transverse section would discover them to be stratified like a roly-poly pudding—only instead of jam and pastry, if their layers were noted on a perpendicular scale, they would range after this fashion: first of all seal, then biped—seal in the center, with biped and seal again at the bottom. Yet, singular enough, these savages are cheerful and really seem to have great capacity for enjoyment. Though in the coldest and most comfortless dens of the earth, they are ever on the grin, whatever befalls them. When they see a white man and his knick-knacks, they grin. They grin when they rub their noses with snow, when they blow their fingers, when they lubricate their hides inside and out with the fat of the seal.

"Truly then, as Sterne says, 'Providence, thou art merciful.'" (p. 98.)

Yes, Providence formed them expressly to live, propagate, and enjoy themselves in that region of ice and snow where no vegetable diet of any kind can be procured, so that their only resource of sustenance lies in the deep-blue sea, drawn from between the cakes of ice which perpetually cover the frozen ocean; and there, encased in bear or seal skins, dwelling in snow houses, and feeding upon the carcasses of fish, seal, and walrus, they laugh and grow fat, and are as happy as the negro is in his mud hut or the ape in his brush habitation, and are much better satisfied with their cold destiny than man is in all the splendor of civilization.

Thus we have a pen-and-ink sketch by a traveler among them, and the only work of their hands described by him is an igloo built of snow, a stone pot, and a covering of seal or bear skin. They feed entirely upon meat, and either eat it raw, entrails, excrement, and all, or pour the excrement into a stone pot and parboil, or rather soak and pickle the meat in it in a tepid state, which is the most that could be done. Thus we discover that the race of Esquimaux are simply a stone pot and a bear or seal-skin garment above the negro in point of intelligent work.

South of the frigid zone, where the climate becomes sufficiently temperate to produce vegetation and sundry wild beasts are able to subsist upon it in the forests and wilderness back from the sea, we find the next higher grade of Indians, whose habitation in summer is a wigwam constructed of poles, set upright in a circle of about sixteen feet diameter at the base, sloping inward to a circle of about three feet diameter at the top, about sixteen feet high, and covered with bark where bark can be had; when not to be had, they cover it with the skins of beasts which they have killed for food, and they excel the Esquimaux in making a bow and arrow; and as household furniture, they substitute a very ingeniously made willow basket for the stone pot of the Esquimaux, in which they cook their chemuck, by heating stones and throwing them into the basket of mash; and some of them who live near the great lakes and rivers construct a very ingenious bark canoe.

Most of the tribes who live in the southern portion of the frigid and northern portion of the temperate zone retain the habit of wearing a sort of blanket covering, composed of some kind of skins; and we are told by travelers that some of these tribes on the Northern Pacific coast live in burrows in the ground, something like the ground squirrels on that coast, which are so troublesome to the wheat-growers of California. Indeed there seems to be as many different species of Indians as there are of squirrels. For instance, there is the species of black, gray, red, ground, flying squirrel and chipmunk, or, as it is sometimes called, striped squirrel, which never cross nor amalgamate; each of which are as true to their species in repeating by propagation the spiritual pattern of the soul of their prototype as the lion is to his through all their generation. So also there is the Esquimaux snow igloo species, the wigwam and skin-covered species, burrowing species, temperate zone species, who go mostly naked; torrid zone species, who go entirely naked; and upon several islands are found one or two species which seem to differ from every other, whose habitations, mood of enjoyment, and mode of working out their destiny of life are as similar as that of squirrels.

We are now writing of the native thoroughbred Indian, with the soul and body of an Indian harmoniously united, as God composed him in the beginning, by just such an arrangement of the simple elements of finite spirit as composed the spiritual pattern of the soul of the prototype of those self-reliant savages who had lived and worked out their destiny of life upon this continent according to their own intuitive desire, will, and personal necessity, with nobody to interfere with them or influence them in any respect, from the time of the flood down to the time of the settlement of the continent by the European nations, with no more change in the form of their wigwam, shape of their bow or arrow, or willow basket, than there had been in the form of a beaver's dam, crow's nest, or card of honey-comb.

We are describing the natural status of that noble savage who met our forefathers in the fifteenth century on the Atlantic coast, and with his naked carcass and bow and arrow battled against the progress of civilization, as the bear and wolf battled and fell back with the same uncompromising hostility—of that wild beastly biped who lived in the same wild, savage, unreplenishing condition that the quadruped beasts had done from the time of their creation down to the discovery of the continent by Americus Vespucius, and the final settlement of the country by our forefathers, and who still persist in the same mode of savage life in spite of the best efforts of conscientious missionaries and ten millions a year appropriated by our Government to civilize and learn them to work. Against all of which they are still battling with the same determination that the bear and wolf battles to annihilation, and who are still found in the western wilderness in their naked thorough-bred condition.

What we are writing here has no reference to the fragments of once powerful tribes of thoroughbred Indians who are now only hybrid abortions of the tribes whose names they bear, who come about one-fourth as near to being Indians as a mule does to being a horse; for a mule is a thorough hybrid, its father being a donkey and its mother a mare; but not so with these abortions, who are called Indians, for their mother was perhaps a half-breed, perhaps a quadroon, perhaps an octoroon, while their fathers are—what shall we call them? not beasts, for a beast would not so disgrace itself. There is but one name for them, and that is men who have by their base lusts transformed themselves into incarnate devils, who would have been stoned to death if the government had done its duty. When we have done with the Indian we will briefly investigate this stock, which is not a species at all, does not come within the classification of any group into which the animal kingdom is divided; and the only thing which this work has to do with them is to draw a line of distinction between them and the native thorough-bred species of animals by whose name they are called. We have made this explanation here in order to have the reader distinctly understand, that when we speak of the Indian races we have no allusion whatever to these hybrid fragmentary relics of the thorough-bred Indian types, but have exclusive reference to the thorough-bred species of Indians which our ancestors found upon this continent when they first settled upon it.

We will also observe here that there were even at that time two groups of hybrids on the continent: one was the Aztecs of Mexico, and the other the Incas' families of Peru, which we will also speak of more fully in another chapter. Still we want the reader to understand that in speaking of the Indian tribes we have no reference to the Mexican Aztecs nor Peruvian Incas.

When God determined to compose the different species of Indians, he considered the kind of constitution that each species would require to live in the different climates; and when he composed the Esquimaux to live in the frigid zone, he made just such an arrangement of the simple elements of finite spirit, forces of animal growth, and celestial substance, in composing the souls of the first pair of Esquimaux, as would make their chief desire and greatest enjoyment of life a snow igloo with a bench of snow inside of it for a bed, their chief glory as a habitation; a stone pot their only cooking utensil and article of household furniture; a garment of seal or bear skin and a blanket of the same material on top of their snow bed the end of their desire for clothing; and plenty of seal, fish, walrus, and bear meat the only food their appetites craved: and that was their eden, and there was where God placed. their prototype in the day that he composed them, male and female created he them, and placed them in that region of perpetual snow and ice, by the side of the sea-shore, where there was plenty of walrus, fish, and seal; where they built their igloo of snow, just like the igloo which Mr. Hall slept in and feasted on raw seal entrails and excrement, as the present crow builds his nest like the nest which the first crow built.

Every species of animal live in a climate most congenial to themselves, and emigrate till they find it, and there they settle and remain. Some emigrate to keep up an even temperature the year round. Cuckoos, for instance, emigrate to keep as near as possible in perpetual spring, and do not remain long enough in any one place to hatch and raise their young, but resort to strategy to perpetuate their species, in doing which they rob some other bird's nest by eating one of their eggs, shell and all, to secrete the theft, and lay one of their eggs in its place, till their litter is completed, and trust the raising of their brood to an adopted mother nurse thus fraudulently obtained, and take their departure to follow up the early blossoms of spring.

The bloom of the early pea is the signal for their departure to a more northerly climate.

The wild geese, too, emigrate to the extreme north to spend the summer, where they can keep in sight of snow mountains, to raise their young in a climate congenial to them, and to the south to spend the winter, not however so far as to be out of the view of snow surroundings. They are a winter bird, and use their strong wings to keep them within the environments of frosty nights, cold storms, and more or less snow. So also every quadruped and biped seeks and finds the climate congenial to its nature, and that is its eden, and there it works out its destiny of life and repeats by propagation the spiritual pattern of its prototype, till the earthly dross of its body is exhausted and it dies, and God directs its course to its everlasting celestial home.

Therefore we know with absolute certainty that every species of wild animal lives in just the climate and exact habitation and mood of life which God composed them to live in; and he who takes them out of it and compels them to live in another climate, feed on other food, environed by different circumstances than those which they choose for themselves when free to act in accordance with their own intuitive desire, sins against them and rebels against the designs of the Almighty, who will bring him to suffer, sooner or later, all that he has caused those animals to suffer.

Consequently the Esquimau now lives where he should live and as he should live, and just where and as God designed him to live; and he who would compel him to live differently sins against God's law and persecutes the Esquimau. And he is endowed with just the degree of intelligence to make his igloo, stone kettle, and skin garments his eden, in the enjoyment of which he grins, laughs, and grows fat, and his cornucopia of enjoyment is full and running over. And precisely the same rule applies to the tribes of Indians who live south of them, partly in the frigid and partly in the temperate zone. They were composed by just such a change in the arrangement of the simple elements of spirit, force, and substance as fixes their status in that climate and makes that their eden home, and there is where God placed the first pair, both of the burrowers and of the wigwam species; and when the burrowers have completed their excavation, and the others constructed their wigwam, and both have constructed their willow baskets, bows and arrows, and a canoe or two for the tribe, and a stone hatchet, their cornucopia is just as running over full of enjoyment as their neighbors, the Esquimaux are in their ice-bound environments.

So also were the tribes of the southern portion of the temperate zone composed by such a change in the arrangement of the simple elements of finite spirit, forces of animal growth, and substance, in composing the spiritual pattern of the soul of their prototype, as gave them a larger and more active personification of the savage tribes, and a constitutional ability to stand even the rigor of the winters of that climate, without even a garment of skins, other than that which God has given them, and just the degree of intelligence which filled their cup of joy to overflowing in the simple wild beastly life which they were living when our forefathers first settled upon this continent, and in which every thorough-bred Indian is living at the present time, notwithstanding all the misspent efforts of missionaries and government appropriations to induce them to live in subordination to the laws of civilization. Their ways are not our ways, neither are our ways agreeable to them. A life of civilization is just as abhorrent to them as the wild savage life they lead is to us. Nothing but force could compel us to live the naked, exposed, beastly life that they lead, which would exterminate us as rapidly as civilization has them. Nothing but force can compel them to live in a mood of civilization; and an experiment of some three hundred years shows that to attempt to live it is the death of them.

The Indians found by the puritans on the east side of the Rocky mountains, between ten and fifty degrees of north latitude, were the largest and most savage specimens of the Indian group, who, on comparison with the Digger Indians, burrowers, and Esquimaux, bear about the same comparative resemblance that the large brown wolves among which they dwell do to the small prairie wolves.

Their inventive working intelligence, however, is on the same key, and their mood of enjoyment is as similar as that of the large and small wolf. When a young Indian has got his bow and arrow he is happy; and when he chooses his mate, their first work is to construct their wigwam, and she her willow basket, and their eden is complete—they desire nothing more. Their fire is built in the center of the wigwam, the hole in the top constitutes the chimney, and as naked as they came into the world they sit and lie about it on the bare ground, tearing to pieces the flesh of the animals on which they feed with their teeth and fingers. Their chemuck consists of pounded grass seed, nuts of various kinds, wet up into a sort of pottage in their willow baskets, and warmed by plunging hot stones into it, around which they all sit in the dirt and scoop it out with their fingers. Some of these tribes who lived near very rich river bottoms, where Indian corn would grow by punching a hole in the ground with a sharp stick or stone, and dropping a few kernels of corn into it, with no other culture or tillage than stepping on the hill to cover the seed, and gathering the nubbins in the fall and burying them in the ground to hide them from the squirrels and other wild

SIMPLE ELEMENTS OF ANIMAL GROWTH. 249

beasts which claimed an equal right to the products of the soil, substituted mouldy corn for nuts in their chemuck.

The only knife or cutting instrument which they had any knowledge of was composed of flint or other hard stone. Their intelligence did not soar into the metal kingdom, neither was there any workers in brass and iron among them. A wigwam dweller in the wilderness—a naked wild beast—whose inventive working intelligence barely reached an octave above the baboon in the universal drama of animal life, still they constitute the highest grade of intelligence in the animal kingdom below the family of man.

The first pair of these Indians which God composed and placed in the country in which they now dwell built their wigwam exactly as the present Indian builds his; made their willow basket, bow and arrow, stone instruments, canoe, and bone spear, which completed their destined work, and they were as happy as a chimpanzee in his brush hut or a monkey on his platform of woven branches in a tree top. And their whole race has repeated the same specific work as constantly as they have the spiritual pattern of their souls.

Therefore, when our pilgrim ancestors disembarked at Plymouth and looked upon the Indian dwellers in that wilderness, and the wigwams, baskets, bows and arrows, stone cutting instruments, canoes, bone spears, and little piles of mouldy corn and acorns buried in the ground, they saw a photograph of the first family of that species as their Creator had composed them, and in their moods the exact mode in which the first family worked out their destiny. The upright poles of their wigwams were of the same length and the same in number, and the diameter of the circle at the bottom and top was precisely the same to an inch, and each strip of bark was laid on in the same way, or if covered with skins, the seams all ran in the same direction; their willow baskets were precisely of the same pattern to the precise number of warp ribs and weaving of each strand; their bows were the same length and the sinews laid upon the back of them the same thickness, and their arrows had the same feather on the hind end and the same flint dagger on the forward point of it; their stone cutting instruments were the same; and their canoes were of the same pattern, because their inventive capacity for manufacturing was limited to that precise number, and the pattern of them was as intuitively fixed in their mind as the form of a card of honey-comb is in the mind of a bee. Take two young Indians (a male and female) from their mother's breast on to a lone island, nurse them at a woman's breast and under her care till they are old enough to gather for themselves, and leave them there, and they will build just such a wigwam, and make just such a basket, bow and arrow, stone cutting instrument, bark canoe, and bone spear, and there their work would end, because the patterns of those things are indelibly photographed in their minds by the spiritual pattern of the soul, beyond which they have no desire. If they had they would make it, for they are just as capable of making anything else as they are what they have made. If they preferred a house to a wigwam, they would make it; if they needed clothing, they would manufacture it; if they needed such food as tillage alone can produce, they certainly would produce it; for they are just as capable of tilling the earth, as far as physical ability is concerned, as we are.

Let him who says it is because they have not been taught, and man does it because he has been taught, bear in mind that man has never had any teacher but his own soul, no instructor but the spirit which God gave him; yet the very first pair commenced the work of civilization just as intuitively as the first pair of Indians made their wigwam and continued to live a wild beastly life. Adam was a gardener, Eve a maker of garments, Cain was a tiller of the earth, and Abel a keeper of sheep: three occupations which no thoroughbred Indian has followed from the day that God made them to the present time. The fact is the Indian is so composed that his wants are exceedingly few, and these he supplies when the means can be gathered at the table of nature, and his simple habitation and other domestic appurtenances are quickly supplied, and he is happy in his wild simple eden; whereas man was so composed that his wants are infinite, and he is in an everlasting excitement, vainly endeavoring to supply them, without the slightest possible chance of ever succeeding. The Indian builds his wigwam and a few other appurtenances, and his soul is at rest. Man completes his marble residence, and he sighs for one constructed of silver; give him a silver mansion, and he will pine for a gold palace; that finished, and he will weep for one composed of diamonds and other precious stones. The Indian gathers a full meal at the table of nature and feasts for a day, then fasts for a week without any inconvenience, and then goes out and gathers again. Man must have his three regular meals a day, works incessantly the year round to fill his garner, and still is in great fear least famine and starvation will overtake him: the reason of which is an equivalent difference in the arrangement of the same simple elements in composing the spiritual pattern of the soul of each species, which no power on earth or in heaven, save God himself, can ever change or alter. The Indian races are destined to live and enjoy their wild beastly mode of life, and do enjoy it, with more contentment than man does or ever will in his ceaseless struggle to learn all that God knows, and vain effort to become equal with him in knowledge.

It will be well for the reader to bear in mind that the constitutional necessity, internal wants, intuitive desires, and intellectual capacity limit the spontaneous actions of all the Indian tribes, in the way of building and manufacturing, to an igloo built of snow, with a bank of snow for a bed, or a burrow under the ground or a wigwam for a habitation; a stone pot or willow basket as furniture for that habitation; a bow and arrow and bone spear as weapons for capturing their fish and game; a hatchet and knives made of stone; garments made of the skins of beasts; and a bark canoe as a useful implement, household furniture, and articles of apparel, besides which they make sundry ornaments for their bodies, such as bone rings for the nose and ears, and strings of bone beads and teeth of various animals and shells of fish to wear about the neck and adorn other portions of their naked bodies. They also make an elder whistle, with which they squeak time to a kind of gutteral grunt and sing-song chorus, somewhat resembling the purring of a cat, together with an intermittent pounding on an untanned hide stretched over a hoop, which constitutes the music to which they dance both their war dances and religious worship. And their only attempt at farming is to punch a little hole in some rich sandy loam, drop a few kernels of corn into the hole, cover it with the foot, and leave it to shift for itself till fall, and then go and gather what few nubbins there may be on it. With this exception their subsistence is gathered from the table of nature. At the time of the discovery of this continent there were at least twenty thousand acres of land for every Indian that was on it. Still they had increased to the fullest extent that they could while feeding at the table of nature. Without tillage the earth produces nothing that man can subsist on, and very little that Indian tribes or any other beast can subsist on. Consequently the country was as sparsely settled by Indians as other beasts when it was discovered by man, notwithstanding they had the entire control of the country for about four thousand years, uninfluenced by man, with a full and free opportunity to develop its resources, without ever having made the slightest effort in that direction; during which time they had partaken of what the earth was able to produce for their sustenance, without replenishing or tillage, with thanksgiving. When there was a succession of plentiful years of wild fruits, nuts, and seeds, there was a slight increase in their number; but when the barren years came, they died of starvation, without ever dreaming that they could increase the products of the earth by replenishing, or provide more abundantly for the increase of their posterity by tillage.

Thus, while these wild beastly bipeds required twenty thousand acres to support one Indian, the Chinese, Japanese, and some other Asiatic nations were supporting an average of from ten to twenty, say an average of fifteen, persons for every square acre they tilled for food, and they had lived just as isolated from the Europeans as the Indians had: so that they never had any instructor but their own souls, no teacher but the spirit which God gave them, yet they supported a population of about three hundred thousand on the same number of square acres, by replenishing and tillage, that it takes to feed one Indian without; and they sustained a far greater number of domestic animals than there were wild beasts feeding in common with the Indians.

Moses is supposed to have copied the following record from Adam's diary, that after every living creature on the face of the earth had been composed and provided for, God looked over it all and saw that it was very good: "and there was not a man to till the ground. So God created man in his own image, in the image of God created he him; male and female created he them. And God blessed them, and God said unto them, Be fruitful and multiply, and replenish the earth, and subdue it: and have dominion over the fish of the sea, and over the fowl of the air, and over every living thing that moveth upon the earth."

Thus we have the key-note on which the intelligence of man was pitched and his nervous system tuned, and that was dominion, replenishing, and tillage: all of which Adam and his family began to do just as intuitively as the Indian races made their wigwams and did their other little work, which no branch of their families has ever ceased to do since Adam and his sons commenced it; neither has any Indian family ever commenced to do any of that work, neither will they ever be induced to do it except by force, because it is entirely outside of the circle of their intelligence, wants and necessities. And two hundred years of vain effort and a useless expenditure of hundreds of millions of dollars for the support of missionaries, Indian agents, and feeding those wild beasts, and making them presents of blankets and trinkets which they have no use for, to say nothing of the expense of supporting an army on the frontier to prevent their murdering and scalping defenceless women and children with their hatchets hid under blankets, both of which were given to them by the Government, ought to satisfy the most philanthropic Indian worshiper that they are an inveterate wild beast, untamable, except per force of bit and spur, and that the only true policy of dealing with them is to make them answer life for life, for every murder perpetrated by them. Point out to them reservations of wild land, while there is any that can be spared for that purpose, and say to them, within these limits you shall be protected, and there you must make your own living, and if ever you leave that reservation you are dead Indians. Everybody that meets you shall kill you, and see that it is executed.

God never designed that the earth should remain a wilderness for wild beasts to feed upon, at the rate of twenty thousand square acres to the individual. That would be a slow way of populating universal space with intelligent souls.

God gave dominion and title to man only in consideration for replenishment and tillage. Therefore, as the Indian has never done either, and has no more capacity for doing them than any other wild beast, consequently he has no more title to land than they, and when man comes and requires it for tillage, he must fall back and keep in line with his wild beastly associates upon the wild land. Why should the products of the calloused hands of the people of the United States be taxed ten millions per year to feed these wild beasts, while they are murdering and scalping the border settlers?

We have deemed it necessary to deal with this subject in a plain straight-forward manner, in order to correct, if possible, some gross errors which have grown out of the absurd idea of the brotherhood of Indians with men, which has led to a vitiating policy on the part of the government in dealing with them, which has caused the most barbarous murders of thousands of defenceless settlers on the frontier, for which the government has paid them in food and clothing, backed up by a false philanthropic cry of a people and press who have the Rocky mountains between them, their wives and children, and the scalping knives of Lo! the poor Indian, which has filled their bosoms of compassion so full, that they have neither eyes nor ears for the cry of the starving poor of their own brethren, among whom are thousands of widows and orphan children of soldiers killed in the late war, defending the life of the government, and go on appropriating from year to year $10,000,000 to feed these murdering savages, who never did a day's work in the way of civilization in their lives, and concerning whom these fulsome philanthropists know no more about their treacherous murdering disposition than they do of the material of which the moon is composed.

The Indian races have no regard for any other species of animal save their own, and suppose they have just the same right to kill men, women, and children that they have to kill any other animal. That, too, is all right as between wild beasts—each animal has to look out for itself; and if one has the temerity to expose itself to the mercy of its carniverous neighbor of another species which is in a hungry mood, he is sure to be gobbled up. Therefore, among all carniverous beasts treachery and strategy are their greatest virtues, as they are their only means of obtaining their necessary food, and extreme caution and constant vigilance are the price of their lives: all of which the Indian races are abundantly supplied with in the composition of the spiritual pattern of their souls, which is all right and proper as between them and other carniverous wild beasts. But for the government of the United States to send out a peace commission, after the Indians have been exercising their wild propensities by murdering, scalping, and robbing the frontier settlers—to make them presents and hire them not to do the like again—is not only a great blunder, but an outrage on the frontier settlers; for it actually amounts to a premium to repeat it. As well might the herdsman feed the wolves that kill his sheep. The only effect would be to call all the wolves in the country to his flock; and the more he fed them the more they would kill, till his last sheep was devoured. Just so it is with the policy which the government has pursued with the Indians. They say, we got well paid for that last massacre: next time we must kill more, and they will pay us more. Both the wolf and the Indian believe that they have a right to kill and eat any animal not of their own species; and so they have of the races of wild beasts, but not of the family of man or his domestic animals. Admit that as to themselves it is right: whenever they do it man has the undoubted right of dominion to subjugate them, and compel them to live in subordination to the laws of civilization, or kill them and put a stop to their depredations. Absolute fear of condign punishment is the only policy that will ever protect the frontier settlers from the scalping knife of the Indian, or their flocks from the depredations of Indians, wolves, and bears. Among themselves all wild beasts have an undoubted right to exercise all their propensities, which never instigate them to kill except for food, and only to the extent of supplying the present want. Their resources for food are the carcasses of other species which dwell among them, and they have sense enough to know that, if they allow the carcasses of the animals they have killed to go to waste, their supply of food would soon run out.

It is not in the nature of an Indian to let the carcass of the animal he has killed go to waste; and the only reason why he does kill men, women, and children in time of peace is because the government pays him for doing it; and whenever his supplies get short, he goes out on a massacring raid, in order to make the government come out with another payment.

Those who have carefully followed the investigation of this group of bipeds through the foregoing chapter will readily understand the arrangement of key-notes in the following scale, appropriated to the part called soprano in the tune of universal harmony, as acted and sung by them in the combined concert of the animal kingdom:

SIMPLE ELEMENTS OF ANIMAL GROWTH. 261

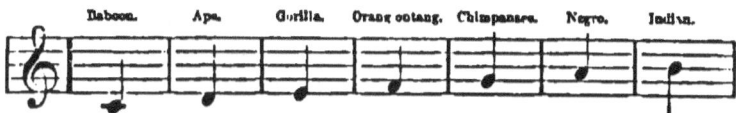

The above scale simply represents the rise and fall of the degrees of intelligence in regular toned grade, from the baboon to the Indian tribes, each of which is just as estimable in the opinion of the Creator as the other, and each acts its part in the drama of life just as acceptably as the other, and with equal satisfaction to themselves. The attentive reader will also readily understand the following arrangement of all the parts in the universal opera of intelligent praise and thanksgiving to God their Creator for composing their spiritual patterns with a degree of intelligence, joy, reasoning capacity, and knowledge of their own thoughts and wants, and ability to supply them:

* * * The above piece represents the seven groups into which the animal kingdom is naturally divided, comprehending every species of animal, from the lowest type of animalcule to the highest type of Indian, which constitutes a perfect anthem of praise and thanksgiving to God, in which there is no discord nor a note of specific intelligence left out, which can be performed either by singing or acting without a discord; *i. e.*, there is not a single species left out of that chain of animal life between the intelligence of the highest grade of Indians and the lowest grade of animalculæ which any change in the simple elements of soul could possibly produce. Consequently no new species could possibly be produced by any process of amalgamation man can instigate among them. And he who attempts to experiment in that line should remember that God hath co nmanded the judges of the people to cause him to be stoned to death, and the final judgment is unquestionably annihilation. For why should God command the judges to use their authority to cut him off from the joys of this life, unless he intends to use his omnipotent authority and decompose his soul back to its native elements and cut him off from the joys of the next? At all events, he who produces a discord in the animal kingdom by perpetrating amalgamation himself, or causes any other animals to do it, takes big chances in annihilation; he sins against the Holy Spirit, by polluting the souls which it made perfect after its kind, filling a note in the scale of intelligent harmony, and produces a beastly abortion, a diabolical discord in the intellectual harmony of the animal kingdom. And Christ testified that said sin against the Holy Ghost should not be forgiven, neither in this world nor that which is to come; and God hath declared that he shall suddenly be destroyed, and that without remedy. The man who imprisons his offspring in the carcass of an Indian or negro, by copulating with a squaw or negro wench, ought to be annihilated and cut off from all enjoyment, both in this world and that which is to come; and we have no doubt that is his final doom. No man can imagine the anguish of that hybrid offspring in this life, whatever its doom may be in the next.

We see many things in this kingdom of animal life which to us appear to be a discord: as, for instance, one species feeding upon another species; but we must bear in mind that this earth is only the breeding matrix of nature, where souls are composed; the great womb of animal formation, in which the great destiny of each species in this whole chain of animals, from the Indian to the animalcule, is to propagate and increase the numbers of its own race in the spiritual pattern of its prototype for the population of its own sphere among the heavenly planets, where they can harmoniously perform their part in the great anthem of praise and thanksgiving to God, and fill their own sphere with intelligent joy, in doing which there is a necessary scramble for sustenance to feed themselves and their little families during this copulating period, not only between species, but also between individual families of the same species, each gathering for its own little household. Therefore it is not expected that all the parts of this universal anthem are to be brought into perfect harmony of tone in this life. Not until they have been extracted from the sandy mold in which their souls are necessarily cast; not until they have escaped from the copulating dross of the terrestrial matter in which their souls are composed, and they have joined the beatific souls of their ancestors in their heavenly sphere, will they be able to act their parts in the anthem in perfect chord with all the parts; not until they have attained to that beatified state of enjoyment, freed from all propagating selfishness, which is so indispensably necessary for the support of a family, will they be able to join in unbroken harmony in singing the everlasting anthem of praise to God for his great mercy and goodness, for composing their souls to enjoy the sphere in which they live with intelligent gratitude to him, and beatific joy and sensations of pleasure to themselves.

But he who will look into this great book of animate harmony, and read the everlasting truths as they stand recorded there in characters so legible and language so plain, that even a child who is willing to be taught can read and understand them, will not see as much discord even in this self-sustaining life of propagation as many have vainly imagined.

From the Indian to the animalcule there is no war of races: no effort to extinguish other species for self-aggrandizement. They all dwell harmoniously together, each species quietly working out its own destiny of propagation in accordance with its own intuitive idea. If there are some carniverous species whose main subsistence is the carcasses of other species, the seizing and killing is quietly done, and only to the extent that their actual necessities require. There is no wanton killing for the sake of destroying the race; they know that their own subsistence depends on the successful increase of the species on which they feed.

The hawk does not seek the nest of its victims and devour a whole brood for a scantier meal than one adult would make, but wisely leaves the young to attain to their full stature, and when their meal-time approaches quietly pounce upon one that is full grown, make their meal of that, and then sail round in common concert, with a kind of patronizing protective influence over the young, something after the manner of a shepherd tending his flocks, who, as necessity requires, occasionally cuts off the head of a very fat one to save his own life. The cat does not seek out the brooding nest and make half of a meal on a whole brood of sucklings, but lies in wait for the adult, quietly seizes him and makes her meal, and leaves the young to acquire their full stature. The massive anaconda snake, having gorged itself, lies perfectly docile for weeks, during which time a rabbit could shelter itself under the shadow of its huge jaws with perfect safety, and a monkey play all sorts of fantastic tricks over its huge body without fear of molestation.

If the lion sounds his war-cry when in great necessity for the supply of his family, it is only for one victim he calls; and having set the frightened animals in motion, he sits down and patiently watches for some frightened victim to rush within his grasp, which he quickly seizes, bears it to his family, and that is the end of the excitement till another meal is required. No hunting for amusement or wanton killing for waste; no war for the extermination of an inoffensive race.

We also find it recorded in this natural history that the grass-feeding gramniverous species of animals increase from ten to a thousand per cent. more rapidly than the carniverous species; so that with all this feeding upon them, they will still fill their heavenly sphere with a much greater number of beatified souls than the carniverous species can.

Those gramniverous, rapidly-propagating species are dull of comprehension, slow of movement, and compose the low notes in the scale of intelligent harmony; while the carniverous species are quick of apprehension, strong, fleet, and fill up the high-key lines of intelligent harmony with just the proper notes to perfect the anthem, which could not be produced on vegetable food, no more than a high-toned instrument could be composed with wires made of lead. These low-toned notes— gramniverous rapidly-increasing species—prepare and refine the food for the high-toned notes. Carniverous species are slow of growth and limited in propagating capacity, which otherwise could not be composed at all, and the best strains of the anthem would be lacking; so that what at first sight appeared to our uneducated senses to be a discord, we find by a thorough reading, comparing, and a proper understanding of the truths recorded in that book of omniscient composing, is in perfect concord with the very nature of things, and therefore with the design of the omniscient composer.

We also learn from this infinite book of God's handiwork, that the souls of all these species, in their beatified state, lose their carniverous propensities; that souls don't eat souls, and that all selfish propensities are shed off with the dross of this propagating body of terrestrial matter, and that in their beatified sphere of existence all the species arrange themselves on the proper line of the anthem, and each individual acts and sings its part in perfect harmony of praise to God and their own beatitude.

www.ingramcontent.com/pod-product-compliance
Lightning Source LLC
Chambersburg PA
CBHW032136230426
43672CB00011B/2349